Special thanks to the authors, editors, art directors,

copy editors, and other staff members of *Fine Homebuilding*

who contributed to the development of the articles in this book.

CONTENTS

ATTICS, DORMERS, AND SKYLIGHTS

FROM THE EDITORS OF **Fine Homebuilding**®

The Taunton Press

The Taunton Press, Inc., 63 South Main Street, PO Box 5506, Newtown, CT 06470-5506

e-mail: tp@taunton.com

Jacket/Cover design: Cathy Cassidy

Interior design: Cathy Cassidy

Layout: Susan Lampe-Wilson

Front Cover Photographer: Bob LaPointe, courtesy *Fine Homebuilding,* © The Taunton Press, Inc.

Back Cover Photographers: (clockwise from top left) Jefferson Kolle, courtesy *Fine Homebuilding,*
© The Taunton Press, Inc.; Andrew Wormer, courtesy *Fine Homebuilding,* © The Taunton Press, Inc.;
Scott Gibson, courtesy *Fine Homebuilding,* © The Taunton Press, Inc.; Scott McBride,
courtesy *Fine Homebuilding,* © The Taunton Press, Inc.

Taunton's For Pros By Pros® and Fine Homebuilding® are trademarks of
The Taunton Press, Inc., registered in the U.S. Patent and Trademark Office.

Library of Congress Cataloging-in-Publication Data

Attics, dormers, and skylights / from the editors of Fine homebuilding.
    p. cm. -- (Taunton's for pros by pros)
  Includes index.
  ISBN 1-56158-779-6
  1. Attics--Remodeling. 2. Skylights. I. Fine homebuilding. II. For pros, by pros
  TH4816.3.A77A8834 2005
  690'.15--dc22
                    2004028673

Printed in the United States of America

10 9 8 7 6 5 4 3 2 1

The following manufacturers/names appearing in *Attics, Dormers, and Skylights* are trademarks:
Andersen®, Band-Aids®, Dacor®, Door-Tite™, Dumpster®, Harrington®, Brass Works, Heat Mirror®,
ImproveNet™, Kohler®, Medex®, Minwax® Jacobean, Plexiglas® ,Q-LON®, Simpson® HD2A,
Southwall Technologies®, Spectrum® Glass, Sun-Tek®, Superglass®, Ultralume™ Lamp,
Velux®-America, Wasco®, Waterworks®

PART 2: DORMERS

PART 3: SKYLIGHTS

INTRODUCTION

Two hundred years ago my tiny Cape Cod house had no dormers or skylights, but it did have an attic. Today the house has three dormers and two skylights, but no attic. That repository of report cards, old love letters, and unused exercise equipment gradually gave way to living space, chiefly bedrooms. And while musty cardboard boxes will occupy the dark, awkward space under the roof without complaint, bedrooms want the headroom, light, and ventilation offered by dormers and skylights.

Unfortunately, colonizing an attic with dormers and skylights poses several challenges, the first of which is design. Too many dormers, especially those added later, have all the architectural grace of a 20-yard Dumpster® deposited on a roof by a tornado. Likewise, most skylights look out of place on historical house styles (that's why my skylights are on the back roof, invisible from the street).

Dormers and skylights are also a challenge to build, in part because they typically involve some complex roof framing. But they're also challenging because they require you to cut holes in a roof, which, generally speaking, is asking for trouble. Hence, flashing and roofing call for meticulous care.

It is my sincerest hope that the articles in this book, collected from back issues of *Fine Homebuilding* magazine, will help you face the challenges posed by attics, dormers, and skylights. Written by builders and architects about their own work, these articles represent the voice of experience—people who have taken a Sawzall to their roof and lived to tell about it.

—Kevin Ireton,
editor-in-chief, *Fine Homebuilding*

Airtight Attic Access

■ BY MIKE GUERTIN

I have been building airtight, energy-efficient homes for more than 10 years. Before that, I built homes I thought were energy efficient. Because I didn't make them airtight, though, I wasn't addressing half of the equation. Making a house airtight means closing the holes between conditioned spaces, such as the living room, and unconditioned spaces, such as the outdoors. I attack big holes first and work my way down until I reach the point of diminishing returns.

The Single Leakiest Hole in Most Houses Is the Attic Access

Often, attic accesses are no more than a piece of painted plywood on a couple of cleats. Most homeowners and builders don't recognize how much conditioned air escapes through an attic access. The access is in the ceiling, probably in a closet, and no one notices the draft. Or they rarely connect the cold drafts on the first floor of the house with the attic access upstairs. However, this hole in the highest part of the ceiling can allow a tremendous amount of heated air to escape into the attic, creating a convection-driven stack effect just like the draft in a chimney. This heated air is replaced by cold outside air, which you might notice as a draft that enters below the front door.

In less than an hour, using just a few materials, some of which ordinarily would end up in the scrap pile, I can build an energy-efficient, somewhat attractive attic access (see the photo on the facing page). It also doubles as a dam to prevent insulation from falling back down into the house whenever the hatch is opened.

Build a Shaft from Sheathing Scraps

I start by making a shaft box that acts as an insulation dam and that supports the hatches (see the drawing on p. 4). In new construction, I size the shaft box to fit between roof trusses, usually 22½ in. wide and about 33 in. long. A couple of 2x4s nailed between the trusses provide nailing for drywall and for the shaft. I make the sides at least 16 in. tall to hold back the attic insulation. For remodeling projects, I size the shaft to fit the existing opening to the attic. If possible, I enlarge small openings to provide easier attic access.

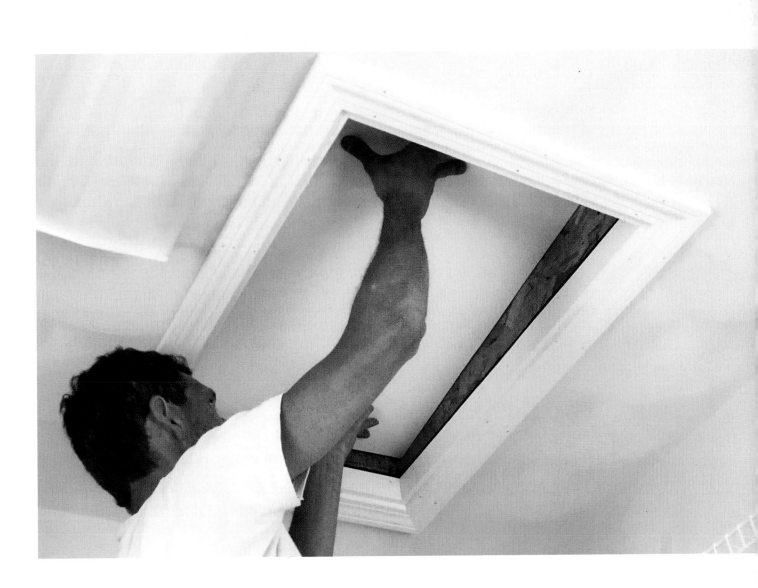

There's no need to make the shaft out of anything fancy, just as long as it's solid. I usually can find several pieces of ½-in. oriented strand board (OSB) or plywood sheathing in the scrap pile that need only to have the edges dressed up and to be cut to size (see the top left photo on p. 4). Even though the shaft fits between framing members, I like to screw the corners together. A little panel adhesive in the joints keeps them tight for a better air seal.

I set the shaft into the rough opening and slide it up until the bottom is flush with the lower edge of the ceiling framing. It then can be screwed or nailed in place (see the bottom left photo on p. 4).

It's important to connect the shaft to the ceiling air retarder. Before raising the shaft into place, I cut an X-pattern in the plastic

vapor barrier that stretched over the rough opening and folded the ears into the attic. After securing the shaft in place, working from the attic I sealed these ears to the sides of the shaft with acoustical sealant. Caulk seals the ceiling drywall to the bottom of the shaft.

Deep Shaft Makes Space for Two Hatches

Making the shaft 16 in. deep gives me the space to install two insulated hatches and provides maneuvering room to remove or replace the bottom hatch. The top panel is larger than the shaft opening and sits on top of the shaft. I make it out of scrap OSB or plywood. On one side of this panel, I glue and screw a 2-in.-thick piece of rigid-foam

Two Insulated Hatches Seal the Attic Access

The shaft is built from sheathing scraps, and the two hatches are built from about $20 worth of foam insulation, MDF, and weatherstripping—not a bad investment in an airtight attic access that might save its cost during the first heating season.

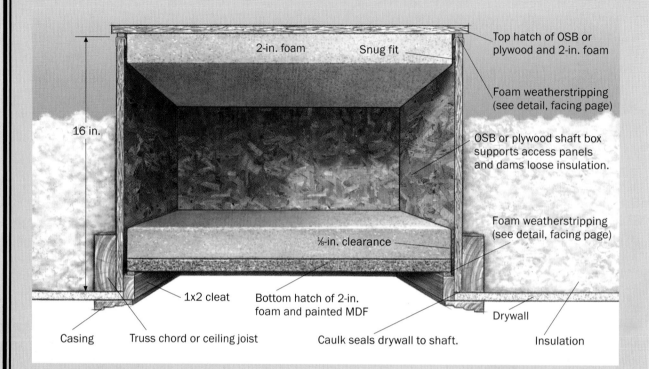

2-in. foam

Snug fit

Top hatch of OSB or plywood and 2-in. foam

Foam weatherstripping (see detail, facing page)

16 in.

OSB or plywood shaft box supports access panels and dams loose insulation.

Foam weatherstripping (see detail, facing page)

⅛-in. clearance

1x2 cleat

Bottom hatch of 2-in. foam and painted MDF

Drywall

Casing

Truss chord or ceiling joist

Caulk seals drywall to shaft.

Insulation

A box made of sheathing scraps forms the attic-access shaft. Drywall screws and construction adhesive hold the sides firmly together (top), sealing against air leaks and making a unit that's easy to lift into place (bottom).

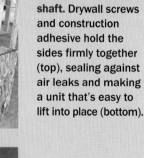

Insulate the hatch. A Pageris gun (Todol Products Inc.; 508-651-3818) dispenses urethane foam, which the author uses to glue foam insulation to the hatch panels.

insulation (see the right photo on the facing page). The foam faces down when the panel is in place, trimmed to fit within the shaft so that the panel is oriented to the shaft opening. You also could add a piece of foam to each side of the top panel.

The bottom panel is visible from inside the house, and even though most attic accesses are in closet ceilings, I like to give the panel a finished appearance. To this end, I use a piece of medium-density fiberboard (MDF) or finish-grade plywood for the panel. I cut the panel about ¼ in. smaller than the inside dimensions of the shaft to make a ⅛-in. space all the way around. A piece of foam the same size as the panel is glued and screwed to its top side.

The bottom panel needs to rest on cleats. I make the cleats from strips of 1x2, then glue and nail them flush with the finished drywall ceiling. To dress up the bottom, I picture-frame the opening with casing.

It's easiest to install the top panel first. A couple of taps on the side of the shaft may be necessary to cause the top panel to drop snugly into place on the gasket. The lower panel needs to be inserted at a slight angle into the shaft to make it past the cleat (see the photo on p. 3).

I've done blower-door tests on numerous homes in which I've installed this type of attic access. Blower doors depressurize the house, and smoke-generating pencils are used to find drafts caused by leaking air. Unlike typical attic accesses, those built my way show no appreciable air leakage.

Price estimates noted are from 2002.

Mike Guertin, *a contributing editor to* Fine Homebuilding, *lives and works in East Greenwich, Rhode Island. He is the co-author of* Precision Framing *(The Taunton Press, 2001), and the author of* Roofing with Asphalt Shingles *(The Taunton Press, 2002).*

Foam Weatherstrip Forms Gaskets

I use ⅜-in.-thick adhesive-backed foam weatherstrip to form gaskets because it's readily available. I stick the foam to the top edge of the shaft and to the top of the cleats. The foam doesn't stick well to the cut plywood or OSB on top of the shaft, so I run aluminum duct tape over the edge first, lapping both the inside and the outside of the shaft box (see the drawing at right). The acrylic adhesive on this tape is relatively long lasting and bonds well to the plywood or OSB. Another method of ensuring a good bond is sanding the edge of the plywood or OSB and brushing on a layer of contact cement. Once dried, the contact cement sticks well to the foam's adhesive. I apply the same foam gasket to the tops of the cleats to form an air seal with the lower panel. Because the tops of the cleats are smooth and clean, extra measures usually aren't called for to get the foam to stick here (see the photo at right).

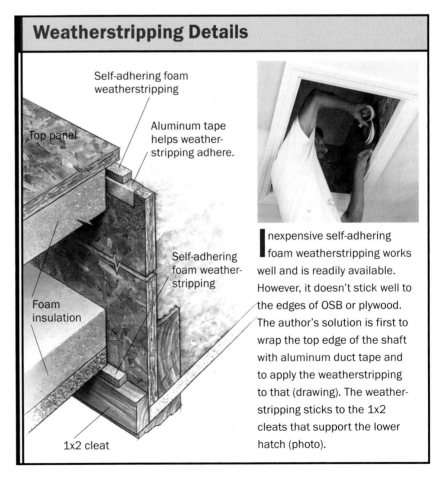

Weatherstripping Details

Self-adhering foam weatherstripping

Top panel

Aluminum tape helps weather-stripping adhere.

Self-adhering foam weather-stripping

Foam insulation

1x2 cleat

Inexpensive self-adhering foam weatherstripping works well and is readily available. However, it doesn't stick well to the edges of OSB or plywood. The author's solution is first to wrap the top edge of the shaft with aluminum duct tape and to apply the weatherstripping to that (drawing). The weather-stripping sticks to the 1x2 cleats that support the lower hatch (photo).

Disappearing Attic Stairways

■ BY WILLIAM T. COX

When I was young and my mother wanted something out of the attic, she would push me up a stepladder and through a little access hole in the ceiling; it was a scary adventure for an 8-year-old, climbing up into a dark, cavelike hole where I thought unknown creatures waited to devour me. What we needed was a disappearing stairway.

Disappearing stairways are available in several styles. All of these stairways have a ceiling-mounted trap door on which the stairway either folds or slides. Nearly all are made of southern yellow pine, although there are a few aluminum disappearing stairways. There are a few commercial models made of aluminum or steel, but this chapter concentrates on residential models.

Sliding stairway. Disappearing stairways are concealed by a spring-loaded ceiling door. Here, the author walks up a sliding stairway made by Bessler with an angle of incline close to that of a permanent stairway.

Folding stairway. Ladderlike sections are hinged like an accordion to the ceiling-mounted door. On this model, made by American Stairways, the treads are painted with bright-colored, rubberized paint.

Disappearing stairways are not considered to be ladders or staircases, and they do not conform to the codes or the standards of either. They have their own standards to which they must conform.

Similar to ladders, disappearing stairways have plenty of labels and warnings to read. On all disappearing stairways there are warnings about weight limits because, inevitably, homeowners fall down stairs while trying to carry too much weight into the attic. Also, labels tell the user to tighten the nuts and bolts of the stairway.

In one stairway manufacturer's literature, the word *safer* was used to describe the fluorescent orange paint used on the stairway's treads. But *safer* was replaced by *high visibility* because one homeowner wore off the paint, slipped, and fell. She sued both the manufacturer and the builder because, she claimed, the treads became unsafe to use.

Stairway companies are constantly testing, upgrading, and improving their products to give the consumer the best, safest, and longest-lasting disappearing stairway possible. And with good reason—over a million units were produced in the United States in 1993.

Folding Stairways

The most popular style of disappearing stairways, folding stairways consist of three ladderlike sections that are hinged together, accordion style. The three sections are attached to a hinged, ceiling-mounted door similar to a trap door. The door and the attached ladderlike sections are held closed to the ceiling by springs on both sides. When you want to access a folding stairway, you pull a cord that is attached to the door and lower the door from the ceiling. The door swings down on a piano hinge. You then grab the two bottom sections of the stairway and pull them toward you, unfolding them (see the right photo on the facing page). When the two bottom sections are completely unfolded, all three sections butt together at their ends, giving strength and stability to the stairway.

Folding attic stairs are measured by the rough opening they occupy and by the floor-to-ceiling height they will service. The smallest folding stairways are 22 in. wide, and they are made to fit between joists 2 ft. on center. These narrow stairways are available in models that will service a ceiling height as short as 7 ft., and there are others that can go as high as 10 ft. 3 in. Keep in mind that a stairway's rough-opening width is appreciably more than the actual width of the ladderlike sections. Because of the attendant jambs, springs, and mounting hardware necessary to operate the stairway, the actual width of the ladderlike section is a lot less than the rough opening. A stairway with a rough opening of 22½ in. is going to have a tread about 13 in. wide.

Folding stairways are rated according to weight capacities; the lightest-rated ones will handle 250 lb., and most of the others have a recommended weight capacity of 300 lb. It is interesting to note that American Stairways, Inc., says in its product literature that you are not supposed to carry anything up or down its stairways. Only an unladen person is supposed to use the stairway. This all sounds somewhat ridiculous to me; it's not as if someone were going to go up into his or her attic crawl space simply to spend a little quality time. The reason why people install disappearing stairways is so they can carry stuff up or down from the attic— Christmas ornaments, baby clothes. However, I tell customers not to carry stuff up the folding stairway. You should have someone hand it up to you. You cannot climb a folding stairway with something in your hands. It's way too steep. I suppose the disclaimer keeps American Stairways out of court if somebody falls down one of the stairways. Also, all folding stairways are for residential use only; a restaurant owner once asked me to install a folding stairway so that he could access a storage area above the kitchen, and I had to refuse.

The real benefit of sliding stairways is their angle of incline.

*It's much easier to cut
and fit (but not nail)
the finish trim while
the stairway is sitting on
the floor in front of you
rather than on the ceiling.*

The smallest folding stairway costs around $75, and the largest, the A-series aluminum folding stairway made by Werner (see Sources below) costs around $211. It fits a rough opening of 2 ft. 1½ in. by 4 ft. 6 in. and accommodates a ceiling height of up to 10 ft. 3 in.

INSTALLING FOLDING STAIRWAYS

Aside from the finish trim, folding stairways come out of the box as a complete, assembled unit. (Other types of stairways require some assembly.) Because most of the installations I do are retrofits into existing buildings, the first thing I must do is cut a hole in the drywall. If possible, I try to mount the stairway alongside an existing joist; this saves some framing work if the stairway is bigger than the space between two joists. Cutting the drywall is not a close-tolerance operation because (within reason) the finish trim will cover any ragged edges. If I have to head off a ceiling joist, I use standard carpentry practices.

Here's a time saver I came upon after installing quite a few stairways. I've found that it's much easier to cut and fit (but not nail) the finish trim while the stairway is sitting on the floor in front of me rather than on the ceiling. Leave ⅛ in. between the edge

Sources

American Stairways, Inc.
3807 Lamar Ave.
Memphis, TN 38118
(901) 795-9200
American makes three models of folding disappearing stairways. The smallest has 1x4 treads and stringers and a rough opening of 22 in. by 4 ft. The largest has 1x6 treads, 1x5 stringers and a rough opening of 2 ft. 6 in. by 5 ft. Scissor hinges join the ladderlike sections. Optional accessories include an R-6 insulated door panel, bright orange rubberized painted treads, and a fire-resistant door panel.

Bessler Stairway Co.
3807 Lamar Ave.
Memphis, TN 38118
(901) 795-9200
Bessler is a division of American Stairways, Inc. Bessler makes a folding stairway as well as a sliding stairway that has a one-piece stringer and slides on guide bars counterbalanced by spring-loaded cables. Bessler's folding stairway has high-quality section hinges that butt when the stairway is opened. Standard features include 1x6 treads and 1x5 stringers, and an R-6 insulated door and bright orange rubberized painted treads.

Memphis Folding Stairs
P. O. Box 820305
Memphis, TN 38182-0305
(800) 231-2349
Memphis makes folding stairs that are very similar to the ones offered by American and Bessler. In fact, a person who worked for Memphis now owns American Stairways. They also sell an aluminum folding stairway as well as a heavy-duty wood model with 2x4 rails and 2x6 treads. Memphis sells a thermal airlock for its stairs that covers the stairway opening. It operates like a roll-top desk and has an R-value of 5.

Precision Stair Corp.
5727 Superior Dr.
Morristown, TN 37814
(800) 225-7814
Precision makes metal folding stairways and a fixed aluminum ship ladder with a 63° angle of incline. The company also makes an electrically operated commercial-grade sliding stair that has a switch at both the top and the bottom of the stairway.

Therma-Dome, Inc.
36 Commerce Circle
Durham, CT 06422
(860) 349-3388
Therma-Dome offers two insulating kits for attic stairs (R-10 and R-13.6) that consist of foil-covered urethane foam boards and touch-fastener tie-downs. These covers seal to the attic floor with a foam gasket. With their high R-values, payback will be quicker in colder climates. The covers cost between $65 and $80. Therma-Dome will fabricate covers for most stairways.

Trico Metal Manufacturing
960 S. Bellevue Blvd.
Memphis, TN 38103
(901) 774-8180
Trico manufactures three different grades of wooden folding stairways.

Werner Co.
93 Werner Road
Greenville, PA 16125-9499
(724) 588-8600
Werner is a large ladder manufacturer that also makes the Attic Master, which is its line of folding stairs. Of particular note is its aluminum stairway with adjustable feet and a load capacity of 300 lb. Options include a wood push/pull rod that takes the place of a pull cord, self-adhesive antislip tread tape and a stairway door R-5.71 insulating kit.

of the door and the jamb. Make sure you mark the location of all four pieces. Once the stairway is installed, you just nail the pieces in place.

Before installing the stairway, I screw two temporary ledgers to the ceiling that project ¾ in. into the rough opening. The ledgers provide a shelf for the stairway's wood frame once I've lifted the unit into the rough opening (see the photo at right). When attaching the ledgers, I make sure they are parallel and that they stick into the opening only ¾ in. Any farther than that, and they might not allow the door to swing open on its piano hinge. Using screws instead of nails to attach the ledger makes it possible to adjust them in case I somehow miscalculate; it also makes them easier to remove when the time comes.

Although some manufacturers warn against it, I usually remove the bottom two ladderlike sections of the stairway before carrying it up the stepladder. Most often I work alone, and some of the stairways are pretty heavy to lift by myself. A 30-in. by 54-in. stairway made by American Stairways weighs 92 lb.

With the ledgers in place, I lift the stairway into the rough opening and set it on the ledgers. Next I carefully open the door fully and center the jamb in the rough opening. Now it's just a matter of shimming the sides of the frame and fastening them to the framing. (Once the unit is installed, I reattach the sections and tighten the nuts and bolts on the hinges with a screwdriver and a socket wrench.)

Most instructions call for nailing the frame in place, but I like to use screws because they are more adjustable than nails, and they are also easier to remove if needed. I start at the hinge end of the stairway jamb. Most folding stairways have a hole drilled at both ends of the piano hinge to screw the hinge into the framing. I always drill another hole through the hinge and sink a third screw (see the top photo on p. 10). I use #10, 3-in. pan-head screws. Adding a

Support during installation. Ledgers screwed to the ceiling provide temporary support for the stairway while the author shims and screws the frame to the rough opening.

third screw can't hurt, and it takes only an extra minute or two.

Instructions call for screwing or nailing into the framing on both sides of the stairway through two of the holes drilled in the arm plate, which is the metal plate to which the door arms are attached. I shim behind the arm plates because it is critical that the arms stay parallel to the ladder and that the pivot plates remain stationary. If they don't, the rivets that hold the arms will wear out from twisting and torquing as the stairway is used.

After I've screwed through the piano hinge and the arm plates, I shut the door and make sure there is an even reveal between the door and the jamb all the way around the door. When this is done I shim and screw off the rest of the wood frame, using #8, 3-in. wood screws.

An extra screw for insurance. A third mounting screw in a folding stairway's piano hinge strengthens the installation. The author drills through the hinge and the jamb and into the rough framing. The large spring at the top of the photo is one of a pair that holds the stairway and its trap door closed to the ceiling.

Accurate cuts are important. A folding stairway that is cut too long puts undue stress on the hinges because the ladderlike sections don't butt at their ends.

CUTTING STAIRS TO LENGTH

Because ceiling heights vary, folding attic stairways come in different lengths, and with the exception of aluminum models, you must cut the bottom ladderlike section to length when installing the stairway. It is not difficult to figure out the cut length, but it is critical to the longevity of the stairway that the length be exact. A stairway that is cut too long will not extend to a straight line, and the ends of the ladderlike sections will not butt together (see the bottom photo at left), putting undue stress on the hinges. And a stairway that is cut too short will stress the arm plates, the counterbalancing springs, and the section hinges.

To cut the bottom ladder section to length, I make sure the arms are fully extended and fold the bottom section underneath the middle section. I rest my leg against the stairway to ensure that it is fully extended, and I take my tape measure and hold it along the top, or front, edge of the middle section (see the photo on p. 11). By extending the end of the tape to the floor (while holding the upper part of the tape against the middle section), I get an exact measurement from the floor to the joint between the two lower sections. I repeat the procedure on the back edge of the stairway to get the length of the back of the cut. Then I remove the lower section, transcribe the measurements, and draw a line between the two points on each leg.

After making my cuts and reattaching the bottom section of the stairway, all that's left to do is unscrew the temporary ledgers from the ceiling and run the precut trim around the frame. I've installed quite a few folding stairways, and I can usually manage to do the whole job in about two hours.

Sliding Stairways

Several companies make folding stairways, but Bessler Stairway Co. also makes a sliding disappearing stairway. Unlike a folding stairway, where the sections are hinged and fold atop one another, the sliding stairway is one long section that slides on guide bars aided by spring-loaded cables mounted in enclosed drums. When the stairway is closed, the single-section stairway extends beyond the rough opening into the floor space above. This is an important consideration because some small attic spaces do not have enough room for the stairway's sliding section.

To access a sliding stairway, you simply pull the door down from the ceiling, similar to the way you'd pull down a folding stairway. Then you grab the single ladderlike section and slide the section toward you, lowering it to the floor. To close the stairway, you slide the single section back up into the opening. A unique cam-operated mechanism locks the ladderlike section in place while you push the door back to the ceiling. A series of spring-loaded, counterbalancing cables makes the door and the ladderlike section feel almost weightless.

The real benefit of sliding stairways is their angle of incline. Folding stairs typically have about a 64° angle of incline. That's pretty steep—more like a ladder than a staircase. Bessler's best sliding stairways have a 53° angle of incline. Sliding stairways, unlike folding stairways, are designed so that the user can walk up into the attic while carrying a load (see the left photo on p. 6).

Sliding stairways are made of knot-free southern yellow pine, and there are four different models from which to choose. The smallest—the model 20—has a rough opening of 2 ft. by 4 ft. and has a suggested load capacity of 400 lb. This model has a stairway width of 17¹⁄₁₆ in. The model 100 requires a rough opening of 2 ft. 6 in. by 5 ft. 6 in. and has a suggested load capacity of 800 lb. The width of the stairway is 18⅞ in. Sliding stairways are measured from floor to floor, rather than from floor to ceiling like folding

Measuring for trimming. With the bottom section folded under the middle section, the author puts his weight against the stairway to ensure it is fully extended. He measures along both the top and bottom edges of the stairway, transcribes the measurements on the bottom section, connects the dots, and makes the cut. The trap door's pull cord can be seen hanging at the top of the photo.

stairways, and the largest, model 100, will service a floor-to-floor height of 12 ft. 10 in. Sliding stairways also have a full-length handrail.

The smallest sliding stairway, model 20 with a maximum ceiling height of 7 ft. 10 in., costs around $225. The model 100 costs around $700.

INSTALLING SLIDING STAIRWAYS

Sliding stairways do not come from the factory as assembled units; installation of these stairways is more for a journeyman carpenter because the finished four-piece jamb is not furnished and must be built on site. Stringers and treads need assembly, and the door and all hardware have to be installed on site.

I frame the rough opening 2 in. larger than the door opening, allowing me to use ¾-in. stock for the jamb and still have ¼ in. of shim space on each side to account for possible framing discrepancies. I rip the jamb stock to a width equal to the joist plus finished ceiling and attic flooring material.

It's possible to attach the finish trim to the jamb while it's still on the floor and then mount the whole unit into the rough opening using braces (called stiff legs or dead men) to hold the jamb to the ceiling while it's being shimmed and nailed. But because I work alone, I screw ledgers to the ceiling the same way I do for folding stairs and then apply the trim later.

I nail the hinge side of the jamb to the rough framing and then hang the door with #10, 1-in. pan-head screws. Next I close the door to fine-tune the opening. After eyeballing the crack along the door edge, I move the jamb in and out to produce an even reveal down each side and then shim and nail the jamb.

Next I lay the stringers on sawhorses and thread the ladder rods with washers through the center holes of both stringers so that the stringers will stand on edge. Ladder rods are threaded rods that go under the wood treads,

giving strength and support to the sections. I install all but the top three treads into the gains (or dadoes) in the stringers, screw the treads to the stringers, then tighten the nuts on the ladder rods. I always peen the ends of the ladder rods to keep the nuts from falling off. It's important to leave out the top three treads so that I can slide the ladderlike section onto the guide-frame bars at the top of the finished jamb.

When I install the guide frames and the two mounting brackets for the drums that contain the springs, I always predrill all of the holes with a %4-in. bit. After 30 years, you would be amazed to see how the wood pulls away from where the screws were put in without predrilling. This causes a minute split to start, and when I repair sliding stairs that are 30 years to 50 years old, the cracks have grown enough that I can stick a finger into them.

Installation of the mounting hardware is pretty straightforward. After putting the stringers onto the guide bars, I attach the cables. *Caution:* I wear gloves and am careful adjusting the cables' tension around the drums. If the cable slips, I could wind up like the old man and the sea, with deep cuts in my hands and no fish dinner.

Another Type of Attic Stairway

Hollywood Wonder Action attic stairways consist of two ladderlike sections mounted to a door in the ceiling. The stairways neither slide nor fold. The mechanical action of the Hollywood stairway is similar to a parallel ruler used by navigators and draftsmen. When the stairway is closed, the bottom section sits on the upper section. After pulling the door down from the ceiling, you lower the bottom section by pulling it toward you, just as you would a folding stair. But rather than unfolding like an accordion, the bottom section pivots on four arms (two on each side of the stringer) and remains parallel

Hinged sections. Hollywood stairways have ladderlike sections that are hinged like a parallel ruler. When opened fully, the sections butt at their ends. On the right you can see the steel-tube handrails. When the stairway is opened fully, the handrails project 18 in. above the attic floor.

to the upper section as you pull on it (see the photo above). When fully extended, the sections butt one another.

Hollywood has five models of stairways. The smallest, model 28-B, has a rough opening of 2 ft. 3 in. by 4 ft. 9½ in. and will accommodate ceiling heights from 8 ft. to 8 ft. 6 in. This model has 6-in.-wide treads 17⅜ in. long. The largest model, the 45-A, has a rough opening of 2 ft. 9 in. by 6 ft. 3½ in. and will accommodate ceiling heights from 11 ft. 1 in. to 12 ft. Model 45-A has 8-in. treads, 23⅜ in. wide.

Hollywood stairways are sold as complete units, requiring only minimal assembly before installation. They are designed for residential and commercial use, and they are the heaviest of the disappearing stairways (115 lb. to 204 lb.) because they have solid ½-in. plywood doors and wide treads and stringers. The treads do not have ladder rods underneath them; rather, they are

mortised into the stringers and fastened with wood screws.

Hollywood stairways have the same angle of incline as sliding stairs; they are just as easy to walk up or down while carrying things. Hollywood stairways have a unique tube-steel handrail that extends 18 in. above the attic floor. This gives the user more support than any other stairway. The stairways can be operated from the top, and this is especially useful for a second-story work-space when you want to pull the stairs up behind you. Another nice feature of these stairways is that they come with mitered trim that's ready to be installed on the jamb. Hollywood stairways cost between $130 and $260, depending on the model.

*Price estimates noted are from 1994.

William T. Cox *is a carpenter in Memphis, Tennessee, who specializes in installing and repairing disappearing stairways.*

Fixing a Cold, Drafty House

■ BY FRED LUGANO

Unintentional "chimneys" let warm air into the attic. The chase containing this plumbing vent pipe extends all the way through the house and should be blocked off to prevent the loss of heated air from the attic.

As a weatherization contractor, I meet a lot of people who are sick and tired of cold, drafty houses. Their problem—and maybe yours, too—is that they live in homes that don't work very well. These houses, new and old, cost too much to heat and to cool. Their paint peels, their roofs dam with ice, and they sometimes make their owners sick. Simply put, these homes lack a good thermal envelope, or an insulated, air-resistant boundary between outside air and conditioned inside air.

Incomplete thermal envelopes are common in old houses, but new ones can have the same or similar problems. Open-web joist systems, cantilevers, balloon-frame walls, and mechanical penetrations allow outside air to penetrate buildings. Sometimes the problem is poorly installed insulation with too many voids; but more often, holes in the building are the real culprits.

Caulks and weatherstripping can help plug small holes (see "Tightening Up Doors and Windows" on p. 18), but this is like using Band-Aids® to treat major wounds. The total area of the holes I'm talking about is measured in square feet, not in square inches. Even so, these problems can now be fixed simply and economically, and buildings a

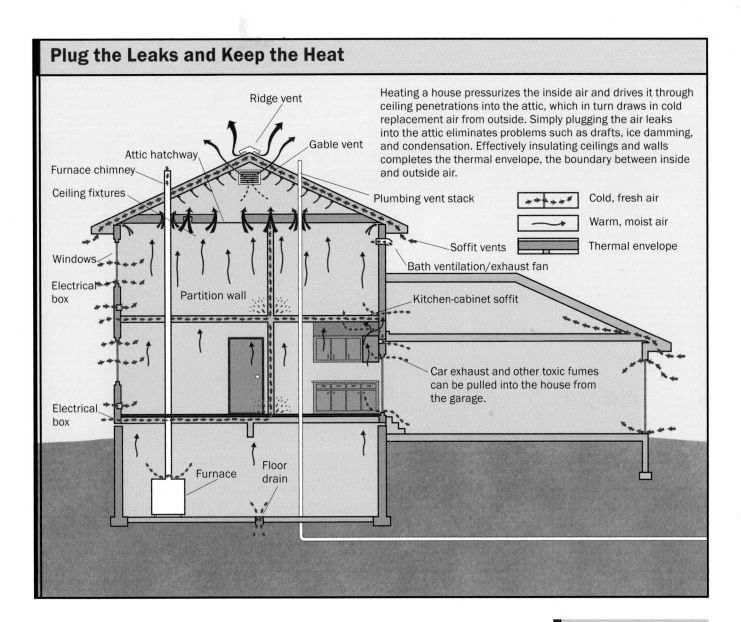

Ridge vent

Attic hatchway

Furnace chimney

Ceiling fixtures

Gable vent

Heating a house pressurizes the inside air and drives it through ceiling penetrations into the attic, which in turn draws in cold replacement air from outside. Simply plugging the air leaks into the attic eliminates problems such as drafts, ice damming, and condensation. Effectively insulating ceilings and walls completes the thermal envelope, the boundary between inside and outside air.

Plumbing vent stack

Cold, fresh air

Warm, moist air

Thermal envelope

Windows

Soffit vents

Electrical box

Bath ventilation/exhaust fan

Partition wall

Kitchen-cabinet soffit

Car exhaust and other toxic fumes can be pulled into the house from the garage.

Electrical box

Furnace

Floor drain

century old can be routinely upgraded to higher performance levels than typical new homes. The principles and methods are also applicable to new construction.

Air Movement in Floors, Walls, and Ceilings Is Bad

Air infiltration is the predominant heat-loss mechanism for most buildings (see the drawing above), so the primary goal of any weatherization effort should be to control air infiltration. Not all infiltration is bad; humans, pets, and furnaces and other combustion devices need a continuous supply of fresh outside air, and the air in most homes should be replaced (either naturally or mechanically) six to eight times per day.

But relying on a home's air leaks is not a good way to provide fresh air. I've worked on buildings that have suffered as many as 30 air changes per day. At that rate, the conditioned air doesn't hang around long enough for the house's insulation to have much of an effect on keeping it in.

Air infiltration is the predominant heat-loss mechanism for most buildings, so the primary goal of any weatherization effort should be to control air infiltration.

Don't seal ducts with duct tape. Fiber-reinforced mastic is a better choice than duct tape for sealing the joints in ductwork because it won't pull away and fall off.

Air infiltration forces warm, moisture-laden air into cold, dry places. The buoyant nature of hot air drives it into every ceiling penetration, and if there are large holes, the house acts like a giant chimney, pulling cold fresh air in from below, heating it, and pouring it into the attic. Loose attic hatches, large cutouts for plumbing vents, exposed beams, and recessed lights are perfect "chimney flues" for these air currents (see the photo on p. 14).

When moist air contacts a cold surface in the wall or attic, water vapor condenses. If the building has a large reservoir of moisture—a wet cellar or an unvented bath, for example—terrible things can start to happen: Recessed lights drool, drywall stains, seam tape lifts, and exterior paint peels off soaked siding and trim. And once the sheathing hits 30% moisture content, mold and mildew can

start growing, a condition carpenter ants and other bugs love. This chimney effect also causes a pressure drop in the basement. Now the living areas are competing with the chimney flues for combustion gases. When the lift through the building overpowers the flues, back-drafting results.

Leaky return ducts in a forced hot-air system (see the photo above) can also vacuum up extra air from the basement and pressurize the living areas, driving conditioned air into the walls. As warm air is forced out, outside air rushes in to replace it. If the basement is tight, that air will come down the chimney, and potentially dangerous backdrafting will start again.

Combustion efficiency will drop, and dangerous pollutants from incomplete combustion, including carbon monoxide, can spill into the basement; be picked up by return ducts; and be delivered efficiently to the rest of the house—a potentially life-threatening situation.

Outside-air intrusion is another classic source of air movement in building cavities, blowing in through openings in walls and running the length of floors before exiting at the other end of the house. This cold outside air immediately comes in contact with warm interior surfaces, chilling them and causing moisture to condense. Ceiling corners are especially susceptible. Here mold can grow, and paint, ceiling texture, and tape can peel off. Contrary to popular belief, most insulation doesn't block air intrusion (see the photo below).

Effective Insulation and Air-Sealing at the Thermal Boundary

A thermally efficient building must have a well-defined boundary between indoors and outdoors. The holes and voids that allow outside-air intrusion are obvious breaks in the thermal boundary, but sometimes it takes a little head-scratching to figure out just where the boundary is. It's a waste of time and money to insulate an area that is actually outside the thermal boundary, so it's important to attack the right combination of floors, walls, and ceilings to yield a complete thermal envelope.

For example, although attics and basements are usually thought of as being transitional areas between inside and outside, they really aren't. There is no "in between" in a properly weatherized house. I generally consider basements and crawl spaces to be inside the thermal envelope because it is difficult to isolate these areas from the living spaces above. Besides, combustion appliances always belong inside, where they operate more efficiently and can contribute Btus to the heated space.

On the other hand, vented attic spaces should always be outside. If the attic is used often, treat the access as an exterior door, and insulate the stairwell walls and under the stairs. When an attic is rarely used, a well-sealed foam hatch over the well is sufficient. I like to use surplus sections of stress-skin panels here. They are heavy enough to compress the gasket we place around the

Dirty fiberglass signals an air leak. As warm air pours through penetrations in the ceiling, dirt is filtered out by fiberglass batts, but the heated air goes into the attic. Holes for wiring should be filled with expanding-polyurethane foam to stop the loss of this air.

Tightening Up Doors and Windows

Unless the glazing is actually broken out, I've found that doors and windows are the most expensive and least productive areas for thermal renovations. A cold house may have between 10 sq. ft. to over 100 sq. ft. of air intrusions and leaks. By contrast, a rattling and ill-fitting door or window won't have more than a few square inches of leakage.

If you do choose to weatherstrip windows, avoid cheap, quick-fix products. They aren't durable or effective enough to warrant the effort to install them. I like Randall's products (Randall Manufacturing Corp. Inc., 200 Sylvan Ave., Newark, NJ 07104; 201-484-7600), which are available in many hardware stores and feature both resilient vinyl bulbs and Q-LON bulbs.

The small, hollow vinyl bulb is easily compressed and conforms well to irregular shapes, which makes it a good choice for attic hatches. I like the Q-LON bulb, often original equipment on new doors, for its ability to conform to large bows and warps by folding and compressing. Metal carriers work well in tight spaces or with metal jambs.

I've also used products from Resource Conservation Technology (2633 N. Calvert St., Baltimore, MD 21218; 410-366-1146) with good success.

Dealing with old doors and windows

Old doors or windows have character and are usually worth fixing up. Simply adjusting the stops will tighten up a rattling double-hung window. Products such as pulley covers (Anderson Pulley Seal, 4232 29th Avenue S., Minneapolis, MN 55406; 612-827-1117), retrofit vinyl jamb liners and side-mounted sash locks (H. B. Ives, 62 Barnes Park N., Wallingford, CT 06492; 203-294-4837) can also help make a tight seal.

Tools of the trade. Counterclockwise from top left: Brown and white Q-LON® weatherstripping; vinyl bulbs mounted in wood and metal carriers; side-mounted sash lock; ratcheting strike plate; brown and white pulley covers with adhesive backing; weatherstrip tape; P-profile EPDM tape.

To make a seal, the bulb must compress slightly when the door is closed, and I've found that the Door-Tite™ ratcheting striker plate (513-891-0210 to order or for dealers) can help folks who have trouble generating enough force to make the door latch completely. This plate has a series of stepped flats instead of one surface for the bolt to catch on.

And what about caulk?

I think life is too short to justify the time spent filling a square foot of $\frac{1}{16}$-in. cracks. If you use caulk, use it sparingly. I carry a dull putty knife and a damp towel to cut corners tightly and to wipe surfaces flush. Rutland makes several paintable, acrylic co-polymer caulks that adhere aggressively and tool cleanly (Rutland Products, P. O. Box 340, Rutland, VT 05702-0340; 800-544-1307). Rutland 500 RTV is the standard when you're air sealing chimneys to sheet metal.

well, they are insulated, and the drywall is ready for paint.

Areas behind a kneewall can fall either inside or outside, depending on the use of the space. Because air infiltration can be a real problem here, special care should be taken to seal off the floor, kneewall, and sloped ceiling from the outdoors. Air sealing and insulating the rafters down to the bottom of the band joist brings this triangular space inside so that it can be used for easily accessible storage. If this space is inaccessible or unusable for storage, my favorite technique is to solidify the entire volume by packing it densely with cellulose, which air seals it and insulates it at the same time. (For more on dense-packing cellulose, read on.)

A Blower Door and Careful Investigation Help to Find the Holes

I use pressure diagnostics to help direct my air-sealing efforts. Depressurizing the inside of a house with a blower door quickly reveals the most significant penetrations of the thermal boundary. But it doesn't take a blower door to find a lot of the major holes in the thermal envelope.

Under natural conditions, pressures are always higher at the ceiling than at the windows. Although wind and mechanically induced pressures are sometimes stronger, hot air applies constant pressure upward toward the ceiling and the attic. As a consequence, ceiling bypasses or holes in the thermal boundary generate more significant natural infiltration through the heating season than do window leaks.

This doesn't mean that door and window weatherstripping isn't cost-effective, but it does mean that most doors and windows don't need replacement. There are reasons to replace windows, but unless there is glass missing or a large gap between sash and jamb, thermal performance is not a compelling one. There are better places to spend energy-conservation dollars.

Another place to concentrate on is the common wall between the house and its attached garage, if there is one. Air leaks here always have the potential to vacuum car exhaust, solvent and weed-killer fumes, and fuel gases into the living space, so this is a spot that requires a NASA-grade air seal. Obvious holes are usually easy to find and fix in the open framing. Caulking framing and sheathing joints down to and along the foundation makes a big difference.

Basements and crawl spaces should also get a thorough inspection. Musty odors are a sure sign that moisture and cold air are mixing and that wood is under attack. Crawl spaces are usually built to save money, and difficult access is often a reliable indicator of potentially significant building defects. We often have to saw our way into crawl spaces, where we can find bare soil, open concrete-block cores, no insulation, no sill seal, empty whiskey bottles, mold, decay, and lots of insect and animal debris.

Blocking moisture in the form of water vapor from the soil with 6-mil poly is an important first step to air sealing here. Cover the ground completely, overlap seams if there are any, and lap the poly right up onto the foundation wall. Then the foundation, sills and band joists can be air sealed and insulated with sheets of rigid foam and plenty of caulk. It's tough to do perfect work in a tight space. Sometimes it's possible to work from the outside by applying rigid-foam panels or stuccoing the stonework.

With the house depressurized by the blower door, I feel for drafts with the back of my hand and spray expanding-urethane foam into trouble spots. Spiders can also offer clues; they always hang their webs in a draft. If the combustion devices have separate fresh-air supplies, foundation walls should be sealed as tightly as possible all the way to the ground, including foundation vents. Although often required by code, foundation vents allow crawl spaces to load up with moisture in warm months and allow cold air to circulate freely through

TIP

It's important always to work at the boundry of the thermal envelope.

the thermal envelope in the cold months. If moisture can be prevented from entering this space, then it doesn't need to be vented out.

It's important always to work at the boundary of the thermal envelope. Often during a blower-door test, an air leak to an electrical outlet, radiator pipe or wainscoting will show up in the middle of the house. But it won't do any good to stop the airflow there because the air will just find another exit point. Leave these interior holes alone and track down where the airflow actually enters the envelope. Air can travel long distances through floor bays and interior partitions in conventionally insulated homes. Air sealing away from the envelope only redirects the airflow to another hole.

Smaller holes and cracks can be filled with caulk or foam. After attic insulation is pulled away, holes can be found and filled. Here the author fills cracks in a plaster-and-lath ceiling.

The Most Effective Air Sealing Is Done in the Attic

Insulation must trap still air; it won't work with air blowing through it. Current residential-insulation practice often ignores this fact. Many homes have vented attics that are actually inside the thermal envelope because the ceiling has so many penetrations. Instead of being trapped by the insulation, warm air pours right up through and into the attic. Everything looks fine when the holes are covered with a blanket of insulation, but when melting snow shows the rafter pattern on the shingles, it becomes obvious that the thermal boundary is really the roof deck. The 12 in. of insulation in the ceiling is yielding an R-value of close to 0.

This isn't the time to add more useless insulation or ugly vents or an ice-dam membrane under the shingles. The best way to correct the problem is to dig through existing insulation and find and seal air leaks (see the photo at left).

Partition walls without top plates in older homes are often a major source of air leaks. Plumbing, wiring, and chimney penetrations should be checked, and light beaming up into the attic from a ceiling fixture below is a sure sign of trouble.

I also look for blackened insulation (see the photo on p. 17). As warm air finds a hole and jets through the insulation and into the attic, the dirt gets filtered out. Batt insulation wasn't designed to stop the loss of warm air from a building, but it does a good job of cleaning it.

Framing around chimneys should be sealed to the masonry with sheet metal and high-temperature silicone caulk. Mechanical penetrations are usually filled with nonexpanding-polyurethane foam from a gun (see the photo on the facing page), while larger holes are best stuffed with fiberglass insulation wrapped in a poly bag (see the photo on p. 22). We generally recycle our empty cellulose bags this way.

For bigger holes, fasten down appropriately sized sheets of rigid foam, metal, or ¼-in. oriented strand board (OSB) and caulk the edges.

There are sometimes areas where it is difficult to rebuild a solid, continuous thermal envelope. For example, suspended ceilings are usually real trouble. When batts are laid over the grid, the assembly behaves like an open skylight, and air flows up through all the openings. In cases such as this, dig into the building until something solid and patchable can be found.

Dense-Packed Cellulose Fills Gaps and Stops Leaks

Once the flow of warm, moist, indoor air is cut off from the attic, the space can be prepared for adequate insulation. Because of the difficulty of installing fiberglass batts properly in an attic, I like to use blown cellulose. Potential ignition sources such as unrated recessed lights and chimneys should be dammed with sheet metal to keep them from contact with insulation. Hatches and soffit vents can be dammed with scrap lumber or plywood. This typically involves cutting and fitting a 10-in.- or 12-in.-deep box, or well, around framing so that it surrounds whatever needs to be dammed and keeps loose insulation out. Flagging electrical-junction boxes is a nice touch that you or your electrician will appreciate when it comes time to find them again. Finally, cellulose can be blown in at low pressure and low speed on top of the existing material to yield an honest R-40 (see the photo on p. 23). Treating open cavities is the easy part of weatherization.

Insulating and air sealing closed cavities is more difficult. In old homes with plaster walls and board sheathing, you can't simply caulk or foam every seam. Fortunately, dense-packing cellulose into closed cavities provides a cheap method of insulating and air sealing in one step; it is so effective that

Ceiling fixtures and plumbing can be trouble spots. Bath fans are a good candidate for caulking, but the irregular hole around the vent stack is better sealed with foam. Insulated ductwork will keep the warm, moist exhaust air from cooling and condensing before exiting.

Larger air passages can be blocked with insulation-filled poly bags. Rafter bays are blocked with bagged fiberglass insulation before being insulated from above with blown cellulose. Later the plaster-and-lath walls will be insulated as well.

it has become a cornerstone of weatherization practice. Instead of being fluffed in, the insulation can be crammed in at twice the conventional density. At 3.5 lb. per cu. ft., it becomes an air-sealing medium too dense for wind to penetrate while maintaining its R-value.

To gain access to the cavities, we drill a 2½-in. access hole from either the inside or the outside (see the photo on p. 24). When we can, we work outside because blowing cellulose is a dusty job. Some contractors drill right through the siding, but usually enough siding can be removed to gain direct access to the sheathing. Later, after the holes are drilled and the cellulose is blown in, the siding can be reinstalled.

A vinyl tube is snaked through until it bumps the end of the cavity. The tube acts as a vertical probe, and if it doesn't extend to either end of the bay, another hole will have to be drilled above or below the blockage. We blow in a lean mixture of air and cellulose at high speed. As the bay pressurizes, the fine cellulose particles flow into every crack.

When the pack becomes airtight, it stalls the flow in the hose, so we pull the tube back until it finds more loose fill. The completed bay is now solid; insulated to R-3.8 per inch; fire, insect, and rodent resistant; and air sealed. When the cellulose is dense packed so tightly that it stops the flow of air from the blower, it will also stop any wind pressure that nature can exert.

We use this method in walls, in cantilevered floors, under attic stairs, and in odd triangular spaces behind kneewalls. Dense pack is also an effective method for insulating cathedral ceilings, the inaccessible thin edges of attics under shed roofs, and inaccessible joist and rafter bays. I think that properly installed cellulose is the finest insulation technique for new construction, too; but its versatility and ability to fill and air seal voids in the wall cavities of old houses makes it indispensable in effective weather-

Blown cellulose makes a good insulation blanket. An alternative to fiberglass batts, cellulose easily flows over and around framing and into cavities. Eliminating gaps or voids helps prevent cold air from washing through the thermal envelope.

ization. Together with attic air sealing, a dense-packed envelope will generally cut natural air infiltration in half before doors, windows, and basement are even touched.

Although dense-packed cellulose won't bring R-19 levels to 2x4 walls, it will still reduce infiltration in ancient buildings to minimal rates. Because most heat loss is caused by air changes anyway, these beautiful and invisibly updated period houses can now perform at higher comfort and efficiency levels than conventionally insulated new ones.

Weatherization Is Tough Business

Insulating a building is physical, dirty work that can take you into tight, uncomfortable spaces. I don't know of anyone who likes

Walls can be packed with cellulose from the inside or the outside. Drilled through either the exterior sheathing or the interior finished wall, 2½-in. holes provide access to the stud bays. A rag placed over the opening while it is being tubed helps control the dust.

working in a confined space at 140°F wearing a respirator.

And what is the payoff for these weatherization efforts? Even in times of relatively stable, low fuel prices, a 20% to 30% return on investment in fuel savings is the norm. Even better are the long-term maintenance issues, such as peeling paint and ice

damming, that effective weatherization helps solve. But best of all is the increased level of comfort for the home's residents: No longer does an old house—or even a new house—have to be cold, drafty, and difficult to heat.

Fred Lugano owns Lake Construction and is a weatherization contractor in Charlotte, Vermont.

Bed Alcove

■ BY TONY SIMMONDS

Tight fit. Into this 9-ft.-long space, the author squeezed a single bed, a row of 30-in.-deep drawers, a bookshelf, and a vanity. A recessed fluorescent fixture illuminates the mirror from above while the vanity table is lit by a lamp behind the mirror. The baseboard reveals the line of the original wall. Above it, drawer fronts, cut from a single 1x10, are screwed from behind to the drawers.

When the middle one of my three daughters grew too old for the loft bed I built for her, the youngest, Genevieve, was happy to inherit it. The loft is in a small bedroom on the second floor of our house in Vancouver, B.C., Canada. Like many second floors of old houses, this one is really a half story, with sloped ceilings where the rafters cut across the intersection of wall and roof. The bedroom has only about 80 sq. ft., so its bed had to be on a raised platform to leave space for a dresser and a desk below.

Soon after she moved into the loft, however, Genevieve started bumping her head on the ceiling over the bed. When she eventually moved the mattress to the floor, I knew it was time for the old bed to go and for a new one to take its place. The bed alcove shown in the photo on p. 25 was the result.

Will It Fit?

The kneewalls that defined the sides of the room had originally been a little over 6 ft. high, leaving a great deal of wasted space behind them. I proposed to recover this space by moving the kneewall over 4 ft. to accommodate a 3-ft.-wide mattress and a bedside shelf beyond that. Given the 12-in-12 pitch of the roof, this would bring the ceiling down below 3 ft. at the new kneewall. Would this be claustrophobic? To answer the question, I mocked up the space with packing crates and plywood to make sure there would be room to sit up in bed. A high ceiling is not a necessity over a bed—within reason, the reverse is true: A lower ceiling increases the sense of shelter and enhances the cavelike quality humans have always favored. Furthermore, a bed in an alcove that can be closed off from the rest of the room has qualities of privacy and quiet that are difficult to achieve in any other way. To get that extra layer of privacy, Genevieve and I decided that her bed alcove should have four sliding shoji screens.

The 9-ft. length of the space would provide room for a dresser and a vanity of some sort, as well as the bed. Drawers underneath the platform would triple the existing storage space. Light and ventilation would come from an operable skylight over the bed.

I had some misgivings about the location of this skylight in spite of the obvious benefits it would confer in terms of light and space. Having never slept directly under one myself, I didn't know whether a skylight so close to a bed would make sleep difficult. But in the end I was seduced by three arguments. First, the skylight would face north and therefore would not be subject to heat-gain problems; second, it would illuminate the shoji from behind; and third, there was the emotional pressure from my client—some drivel about the stars and the treetops and falling asleep to the sound of rain on the glass.

Tight Layout

Juggling existing conditions is the challenge of remodeling. None can be considered in isolation. For example, I had to decide whether or not to keep the existing 7-in.-high baseboard. I could have moved it, but I wanted to leave it in place, partly for continuity and partly to avoid as much refinishing as possible. Starting the drawers above the baseboard also meant that the baseboard heater already on the adjoining wall wouldn't have to be moved to provide clearance for the end drawer.

Four drawers fit into the space between the baseboard and the mattress platform. The drawers are 7 in. deep (6½ in. inside), which is ample for all but the bulkiest items. This brings the mattress platform to a height of about 18 in. With a 4-in.-thick mattress on top of it, the bed still ends up at a comfortable sitting height.

In plan, the mattress takes up almost exactly three-quarters of the 9-ft.-long space. The leftover corner accommodates a makeup table with mirror above and more

Bed Alcove Anatomy

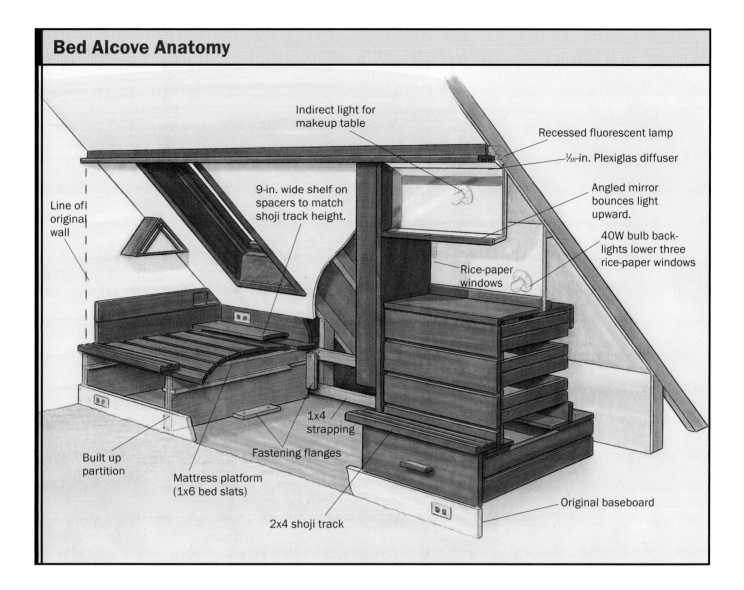

Indirect light for makeup table

Recessed fluorescent lamp

1/16-in. Plexiglas diffuser

Angled mirror bounces light upward.

40W bulb back-lights lower three rice-paper windows

9-in. wide shelf on spacers to match shoji track height.

Rice-paper windows

Line of original wall

Built up partition

Mattress platform (1x6 bed slats)

Fastening flanges

1x4 strapping

2x4 shoji track

Original baseboard

drawers below. I imagined that the shojis would draw a discreet curtain over the wreckage of eyeliners, lipsticks, mousse, and everything else that was supposed to go in the drawers but never would.

I knew that this vanity area, and especially the mirror, would need to be lit, but beyond making sure there was a wire up there somewhere, I didn't work out the details during the preliminary planning. I was in my fast-track frame of mind at this stage of the project.

Site-Built Cabinet

The underframe of the bed is a large, deep drawer cabinet. You could have it built by a custom shop while you get on with framing,

wiring, and drywalling. Custom cabinets are expensive, though, and after nearly 10 years in the business of building them, I appreciate the virtues of their old-fashioned predecessor, the model A, site-built version. It's economical in terms of material and expense, and you can usually get a closer fit to the available space.

The partitions supporting my daughter's bed are made from ⅜-in. plywood sheathing left over from a framing job (the rewards of parsimony). Each partition is made from three layers of sheathing (see the drawing on p. 28). The center layer runs the full height of the partition, but the outer ones are cut in two, with the drawer guide sandwiched between top and bottom pieces.

The guide is simply a piece of smooth, fairly hard wood, ¾ in. thick and wide enough so that it projects ⅜ in. into the drawer space.

Unless circumstances demand the use of mechanical drawer slides, I prefer to hang drawers on wooden guides. I have provoked derision from cabinetmakers because I use wooden guides in kitchens, but when it comes to bedrooms I am almost inflexible. Even large drawers like these will run smoothly year after year if they are properly fitted and if the guides are securely mounted. And for me there is a subtle but important difference between the sound and feel of wood on wood and that of even the finest ball bearings.

I attach the guides with screws rather than with glue and nails so that they can be removed, planed, and even replaced without difficulty should the need arise. A groove in the partition to house them is not necessary, but it's a way of ensuring that they all end up straight and exactly where you want them.

For this job, the pairs of guides on the three middle partitions had to be screwed to one another, right through the core plywood. I drilled and counterbored all the screws and clamped the partition to my workbench to make sure everything stayed tight while I drove the screws. Then, with the partition still on the bench and after inspecting every screw head carefully for depth below the surface, I set the power plane for the lightest possible cut and made three passes over each guide: first over the back third only, then over the back two-thirds and, finally, over the whole length of the guide. Tapering the guides so that they are a fraction farther apart in the back allows the drawer to let go, rather than tighten up, as it slides home.

Partition Alignment

Installing the partitions is the trickiest part of a site-built cabinet job like this one. I said earlier that you could save on materials by building the cabinet in place, but you can't save on time. After all, anyone with a table saw can build a square cabinet in the shop, but building one accurately in a closet or in an unfinished space under the rafters takes patience and thoroughness. The key to success is to establish a datum line, then lay out everything from this line, leaving the wedges of leftover space around the perimeters to be shimmed, trimmed, fudged, and covered up as necessary.

In Genevieve's room, the existing baseboard provided a datum line in both horizontal and vertical planes. First, I divided the baseboard's length so that the four drawer fronts would lie directly below the shoji screens. I ran one screw into each supporting partition, about 1 in. below the top edge of the baseboard. Then I plumbed the front edge of the partition and secured it with a second screw near the bottom of the baseboard. With the front edges located and the partitions standing straight, the next job was to align them to create parallel, square openings.

I built the new kneewalls 48¾ in. back from the inside face of the baseboard. This allowed me to run a couple of 1x4 straps horizontally across the studs to provide anchoring surfaces for the 48-in. partitions (see the drawing on p. 27).

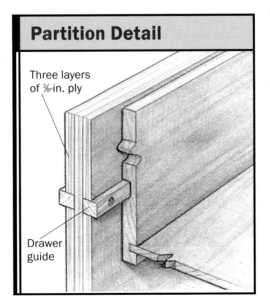

Partition Detail

Three layers of ⅜-in. ply

Drawer guide

Aligning partitions. Load-bearing partitions made of three layers of ⅜-in. plywood separate the drawer bays under the bed and support the mattress platform. The drawer guides are sandwiched between the outer layers of plywood. The photo shows the hardboard panels that helped align the partitions. Once the panels were in place, the partitions were screwed first to the baseboard and then to strapping along the stud wall. The 1x2s clamped to the leading edges of the panels are gauges that will be used to determine the depth of the grooves in the drawer sides.

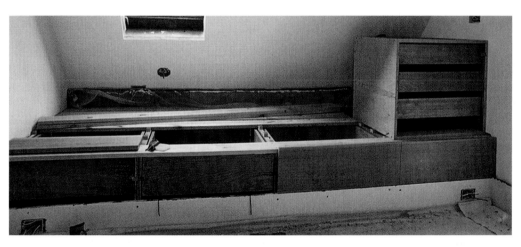

Linked by slat. The partitions are tied to one another across their tops by 1x6 pine slats. Spaces between the slats provide ventilation for the mattress. At the right side, the carcase for the vanity drawers sits directly atop the bottom drawer partitions.

To align the partitions, I used hardboard cut to the full opening width (see the top photo above). As long as the hardboard is cut square, and the partitions are secured so that the hardboard fits snugly between them, the resulting opening will also be square. I used screws to fasten the plywood flanges that held my partitions in place, just in case adjustment should be necessary.

When all the partitions were in place, I cut pieces of 1x2 to the exact dimension between each pair of drawer guides. Centered on the drawer fronts, the 1x2s are gauges that show how deep the grooves need to be in the drawer sides.

The bed slats also act as ties to link all the partitions together (see the bottom photo above). I used dry 1x6 shelving pine for the slats, but almost anything that will span the distance between supports will do. I left an inch between the slats to keep the mattress well aired. I learned this the hard way when an early bed I built on a solid plywood platform developed mildew on the underside of the mattress cover.

Fitting the Drawers

Before putting anything on top of the platform, I built and fitted the drawers. The

drawers have a ⅛-in. clearance between their sides and the partitions. The ⅜-in. projection of the drawer guide thus creates a ¼-in. interlock with the sides. All the drawers are 30 in. deep, but I let the sides extend 6 in. beyond the back of the drawer. The extensions support a drawer right up to the point at which its back comes into view.

If time and budget allow, I use a router jig to dovetail the front of a drawer to its sides, but the back just has tongues cut on each end that are glued and nailed into dadoes in the sides (I take care not to put any nails where the groove for the guides will be plowed out). The drawer bottom rides freely in a groove cut in the front and the sides and is nailed into the bottom edge of the back, which is only as wide as the inside height of the drawer. Fastening the bottom here helps keep the drawer square.

Fitting the drawers should present few problems if they are built square and true and if time and care have been invested in positioning the partitions. Don't try for too tight a fit, especially in the width of the groove. My guides were ¾-in. material, and I plowed out a ¹³⁄₁₆-in. dado in the drawer side. They're not sloppy.

On the other hand, you should be more stingy about the depth of the grooves. Remember, the guides have been planed to allow increasing clearance as the drawer slides home. Too much slop here can cause the drawer to bang about from side to side and actually hang up on the diagonal. You can always plow a groove out a little deeper. A router with a fence or a guide attached is the ideal tool for this because you can easily make very small adjustments. If things go wrong, you can glue a length of wood veneer tape into the dado, but it's nicer not to have to do that.

I dress the groove with paraffin wax, but only when I'm sure the drawer doesn't bind. Patience in working toward a fit has its reward here. The moment that a wood drawer on wood guides just slides into its opening and fetches up against its stop, expelling a little puff of air from the cabinet, is a moment that provides much satisfaction.

Beyond the Footboard

With the drawers and the platform in, I had to decide what to do about the divider between the bed and the vanity. Here was where the self-imposed constraint of using the existing baseboard as the perimeter of the alcove began to bite. Because its height was determined by the slope of the ceiling, the mirror over the dressing table had to be as far forward as possible. But to bring it right up against the inside edge of the upper shoji track would eliminate the space required for a light above the mirror. And even that would put the top of the mirror at barely 6 ft. Temporarily derailed on the fast track, I tried to find other ways to light the mirror and kept coming back to the necessity of recessing a fluorescent fixture into the ceiling.

The fixture I used is a standard T-12 fluorescent fixture equipped with an Ultra-lume™ Lamp (Phillips Lighting Co.; 800-555-0050). The lamp emits more lumens per

Bookcase wall. Shelves deep enough for paperbacks are affixed to a ¾-in. birch plywood panel between the bed and the vanity. The squares at the end of each shelf frame rice-paper windows that are backlit by bulbs behind the vanity drawers.

Headboard. A reading light inspired by the bookcase's square-and-triangle motif lights up the headboard side of the bed. On the left, wooden tracks for shoji screens frame the alcove.

watt than a standard cool-white lamp and has a higher Color Rendering Index, both important factors in getting an accurate reading on colors, such as those at a make-up table.

Casting an even light across the face of the person standing at the mirror is necessary. So I put a narrow strip of mirror along the bottom edge of the large mirror, angled upward to bounce the light where it can fill in shadows.

Bookcase Wall

As for the partition between dressing table and bed, my fast-track conviction that it could not be frame and drywall held up better. My daughter wanted more bookshelves, and the foot of the bed was a logical place to put them (see the photo on the facing page). I made the back of the bookcase out of ¾-in. birch plywood, which could be finished naturally on the book side and painted white on the dressing-table side to look like a wall.

To light the makeup table, I mounted a standard incandescent ceiling fixture in the space behind the mirror. On a playful impulse, I wired another of these lower on the sloped ceiling in the space behind the vanity drawer (the case for these has no back, so the fixture is easily accessible). Then, after carefully laying out the location of the bookshelf dividers and following a square-and-triangle motif suggested by the conjunction of the ceiling and the shelves, I jigsawed the holes in the birch ply and glued rice paper over them. This created little backlit rice-paper windows in the bookshelves. The dividers cover the edges of the paper. The only slight snag in this assembly is that the plywood thickness causes a shadow line, which can be seen where the backlighting travels at an angle through the window. If I'd thought of it in time, I could have easily eliminated the shadows by beveling these edges with a router.

The bedside reading light was more of a problem. Initially, I placed my standard ceiling fixture under the skylight as far down the slope of the ceiling as I could. I made a cardboard mock-up of the rice-paper shade that I had in mind to establish just how big it should be—the trade-offs being the height of the fixture, the size of the shade, and its proximity to errant elbows. I thought I had a satisfactory balance, so I went ahead and made the lamp. But Genevieve put her elbow through it the first night she slept in the bed. I forgave her and accepted the lesson. The second reading light ended up above the head of the bed (see the photo on p. 31).

What about the Shojis?

The shoji screens have yet to be made, and it now seems unlikely they ever will be. Although she was initially keen to have them, Genevieve now believes they would get in the way, and I agree with her. We analyzed the patterns of opening and closing that might be required during a typical day and night. It became clear that in spite of the desirability of drawing a curtain over the unmade bed by day and the unfinished homework by night, this teenager would rather live and sleep in one room—at least for the time being—than be bothered sliding screens to-and-fro all the time. A feeling of confinement was also a factor. Having tried out the bed myself one night when she was sleeping at a friend's house, I too felt I might want more distance between myself and any enclosing screen.

I admit that this was something of a blow to my vision of the room. What about the function of the skylight as a backlight for the shoji? What about the square-and-triangle motif I was going to incorporate into the shoji lattice? Ah, well, at least I hadn't made them already. And the grooves in the bottom track appear to work perfectly as 9-ft. long pencil trays.

The rejected shojis and the difficulties I had with the makeup light and the height of the mirror were all results of my decision to keep the bed alcove within the area beyond the existing kneewall. If I had moved this line 6 in. to 12 in. into the room, I could have raised the upper shoji track a few inches, creating plenty of space to mount the mirror light, the reading light, and the shoji screens. The amount by which this would have reduced the size of the room would have been insignificant in relation to the space gained by building in the bed and the dressing table—a case of choosing the wrong existing condition to work from.

At least the client is satisfied. The project was completed during one of the long dry spells that Vancouver is famous for. Finally, one morning when the spiderwebs were glittering and the earth smelled refreshed and autumnal, Genevieve appeared downstairs with a beatific smile on her face. "It rained on my skylight last night," she said.

Tony Simmonds is a designer and builder in Vancouver, British Columbia, Canada.

A Fresh Look for an Attic Bath

■ BY JACK BURNETT-STUART AND JULIA STRICKLAND

Skylight sheds light on a dim bath. Reflected light from a shower skylight dances on white-tiled walls, filling this attic bathroom with natural light. Photo taken at A on the floor plan (p. 34).

The designer of this attic conversion had sure run out of inspiration when it came to planning the bathroom. Built in 1906, the upstairs of this South Pasadena bungalow was converted to living space in the 1970s. The bathroom wasn't small, but with just a single corner window, it was short on natural light. And coupled with a dismal brown-and-tan color scheme, the room was somewhat less than cheery.

Furthermore, the undersize tub and a claustrophobic shower had been inexplicably crammed into the space under the sloping attic ceiling (see the before floor plan below).

Our clients, Karen and Ken, wanted a contemporary bathroom that would complement the Craftsman style of the original house without being slavishly historical. Both Karen and Ken work in set decoration

Dramatic Changes within the Confines of Existing Plumbing

Keeping plumbing changes to a minimum kept construction costs and hassles down while allowing more resources to go toward finish materials. Turning the tub 90° made room for a full-size tub as well as a generous shower platform. The skylight over the shower lets in abundant natural light and creates additional headroom for the shower.

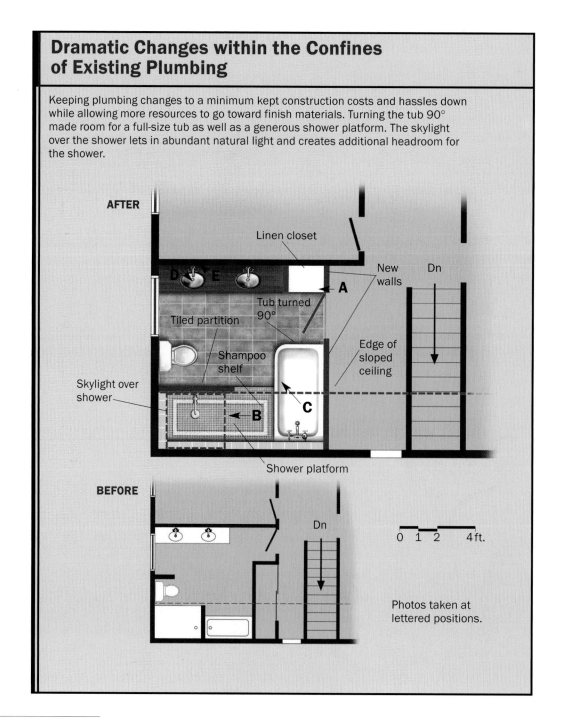

in the film industry, so they were able to give us clear ideas about the palette of finish materials that we eventually used for their bathroom (see the photo on p. 33).

Killing Two Birds with One Skylight

Our first priority was to bring in more light and air. Next, the shower area needed a taller ceiling so that Ken, who is well over 6 ft. tall, could at least stand up straight while showering. As with many bathroom remodels, we had to work with the existing plumbing layout, so moving the shower to another part of the room wasn't an option.

Our solution was to eliminate the old enclosure and to create a doorless shower separated from the rest of the room by a tiled partition. The new shower area is top lighted and ventilated by a large operable skylight, which increases the headroom in the shower enough for Ken to stand upright (see the top photo at right). A shower head also rises into the space gained under the skylight.

We replaced the old tub with a full-size recycled cast-iron tub turned 90° (see the bottom photo at right). The new tub merges with the platform that forms the shower floor. This arrangement let us omit a shower door or curtain.

White Tile Reflects Added Light

Ken and Karen wanted the subtle earth tones of unpolished travertine for a nonslip floor in the bathroom. They chose large rectangular tile for the dry area of the floor and 1-in. tumbled travertine mosaics in several colors for the shower platform. The travertine step that leads up to the platform and continues around the tub was perfectly executed by tile installer Joe Stanley.

Standing room only. The extra space under the skylight provides headroom for a tall person to shower in comfort. Photo taken at B on the floor plan (facing page).

Recycled tub built into the shower platform. The tub is built into a raised platform that forms a wet zone draining back to the shower. The tile wainscot extends up for a shampoo shelf above the tub. Photo taken at D on the floor plan (facing page).

Vanity inspired by history. Having seen the butler's pantry in the nearby Gamble House, the owners opted for polished-metal sinks and a wooden countertop made of tough, stain-resistant goncolo alves. Photo taken at E on the floor plan (p. 34).

To take full advantage of the increased light, the walls were finished in white crackle-glazed subway tile. In the shower, the white tile extends up 5 ft. and then caps the wall to create a high shelf for toiletries. The rest of the bathroom has subway-tile wainscot up to windowsill height. A unifying line of relief-trim tile wraps the whole room.

Vanity Inspired by the Greene Brothers

Inspired by the butler's pantry in Greene and Greene's nearby Gamble House, Ken and Karen wanted to use metal sinks in a wooden counter for the bathroom vanity (see the photo above). They chose goncolo alves, a dense hardwood capable of resisting an onslaught of soap, water, and toothpaste.

This wood is somewhat like teak, and its reddish tone and pronounced grain provide a warm contrast to the hard surfaces of floor and wall. But working with the wood proved to be a headache for contractor Ben Harrison. The wood's wavy grain and hardness caused it to chip when it was routed. Nickel-finished hardware and sinks as well as matching exposed plumbing below the counter (see the photo below) round out the clean, utilitarian yet decorative effect that our clients had appreciated at the Gamble House.

Both the door and window were made new. They were glazed with a single pane of narrow-reed glass and trimmed to match the 1906 details elsewhere in the house. Paneled doors on the linen closet and medicine cabinet were painted to match the wood trim.

In subtle contrast to these traditional details, Ken and Karen chose contemporary glass sconces beside the mirror and above the tub, and recessed low-voltage downlights in the ceiling. The result is traditional yet idiosyncratic, a bathroom that looks fresh and modern but still fits in with the old house.

Jack Burnett-Stuart and Julia Strickland *have an architectural practice in Los Angeles, California.*

Sources

Vanity sinks
Bates and Bates
Garland MBO1318
(custom nickel-plated by owners)
(800) 726-7680
batesandbates.com

Faucets
Harrington® Brass Works
Vanity: Victorian 30-200
Tub: Victorian 20-306
(201) 818-1300
harringtonbrassworks.com

Shower faucet/head
Waterworks®
Norfolk RUSV67
(800) 998-2284
waterworks.com

Decorative plumbing. The exposed drains under the vanity are attractive and utilitarian. Photo taken at C on the floor plan (p. 34).

Adding On, but Staying Small

■ BY HARRY N. PHARR

The word *cottage* used to suggest small-ness. No longer is this true. Today, a country cottage can be a 4,000-sq.-ft. second home with a three-car garage and a lap pool. This cottage, however, lives up to the humble origins of its name. It is a small country dwelling north of New York City, built mid-century as a summer cabin and, alas, not built very well.

The house started life as a one-story cabin with a gable roof and three small rooms with no hallways between them—you had to pass through one room to get to another. At some point, a bedroom was added to the back of the house under a shed roof, making the layout that much more clumsy.

The current owner, my client, bought it as a summer home, too. But after years of forgivable neglect, the little cottage could not even live up to this modest task. It was truly tiny—even discounting for today's gargantuan standards—and flimsy. On my first visit, the floor joists deflected under my weight.

Despite its shortcomings, the cottage was worth saving, and saving in more than spirit alone. The humble structure sat near the top of a slope accessible only by a footpath. It overlooked a clearing that my client had,

A cottage grows in size but not in scale. Built in the 1940s, this three-room house had quirky charm (above). To gain space, the first floor was enlarged, and a low-profile second story was added (facing page). Photos taken at A on the floor plan (p. 40).

over the years, transformed into a magnificent English-style garden.

My client, Berenice, is a gardener and a bicyclist (she does not own a car), and she did not want to destroy this intimacy. There could be no driveway, no massive deck, no towering two-story addition. Was there a way, she asked, to carve out more space without upsetting the balance between house and garden? Was there a way to grow while keeping the cottage a cottage?

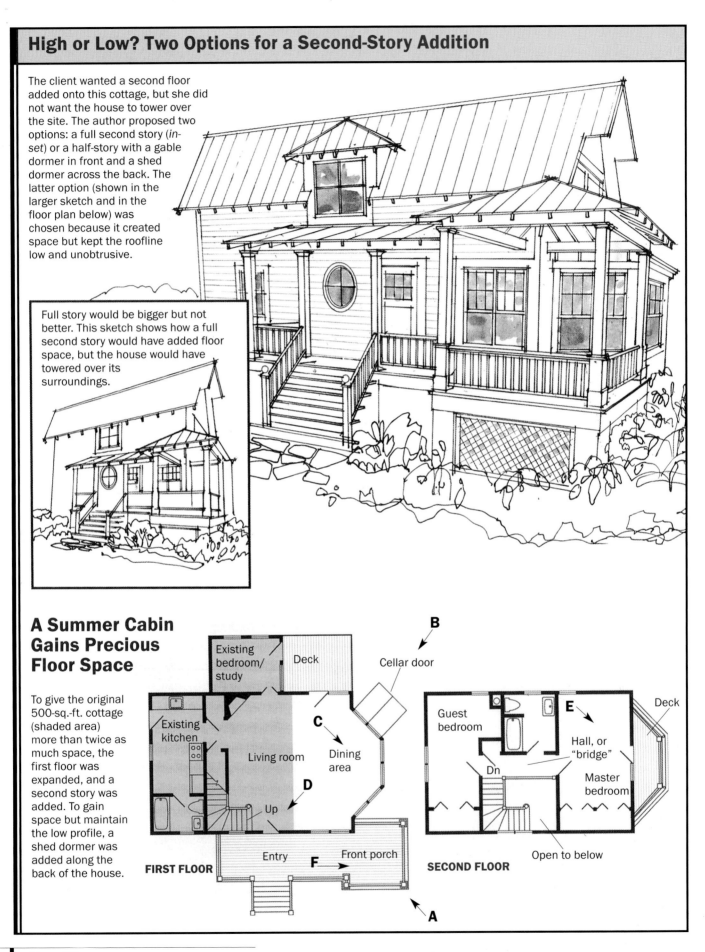

The client wanted a second floor added onto this cottage, but she did not want the house to tower over the site. The author proposed two options: a full second story (*inset*) or a half-story with a gable dormer in front and a shed dormer across the back. The latter option (shown in the larger sketch and in the floor plan below) was chosen because it created space but kept the roofline low and unobtrusive.

Full story would be bigger but not better. This sketch shows how a full second story would have added floor space, but the house would have towered over its surroundings.

A Summer Cabin Gains Precious Floor Space

To give the original 500-sq.-ft. cottage (shaded area) more than twice as much space, the first floor was expanded, and a second story was added. To gain space but maintain the low profile, a shed dormer was added along the back of the house.

B

Cellar door

Existing bedroom/ study

Deck

Existing kitchen

C

Living room

Dining area

D

Up

Entry **F** Front porch

FIRST FLOOR

A

Guest bedroom

E

Deck

Hall, or "bridge"

Dn

Master bedroom

Open to below

SECOND FLOOR

A long dormer expands the upstairs. The low roof limited the upstairs floor space, so a long shed dormer was added along the back. The overhanging rake shelters the deck and makes the roofline more interesting. Photo taken at B on the floor plan (facing page).

A Summer Cabin Short on Amenities

Big or small, it seemed clear that we were looking at a major renovation and an addition. Berenice wanted a large, open living room with a fireplace and views of the garden. She also wanted a master bedroom and guest bedroom, along with a conveniently placed second bathroom (the original bathroom remained off the kitchen). In short, she wanted space.

Along with adding more space, we would have to upgrade the existing structure and systems. A summer cabin can get by with things a year-round house cannot, and this house had barely gotten by. It was built without heat and remained that way until

someone installed electric baseboards. Most of the house sat on concrete piers over a 30-in. crawl space, except for an 8-ft. by 10-ft. basement underneath the kitchen and bathroom. This cinder-block basement was prone to flooding, and as a result, the water heater was rusted beyond repair. The plumbing and electrical systems were poor, the insulation was less than adequate, and the framing was lightweight. The 2x6 floor joists were on 20-in. and 24-in. centers, accounting for the bounciness of the floor.

So how do you gain space without destroying what made the property special in the first place? The obvious answer was to expand the first floor and to add a second story for the two bedrooms and bathroom.

Hiding a Gutter behind a Site-Built Wall

Instead of off-the-shelf gutters on the second-floor deck, the architect and builder made the gutter an integral part of the roof detailing. The roofing—a single-ply, torch-down waterproof membrane—wraps into a trough behind the molding. The decking and sleeper assembly, or duckboard, sits on top of the membrane and can be removed for cleaning leaves and other debris.

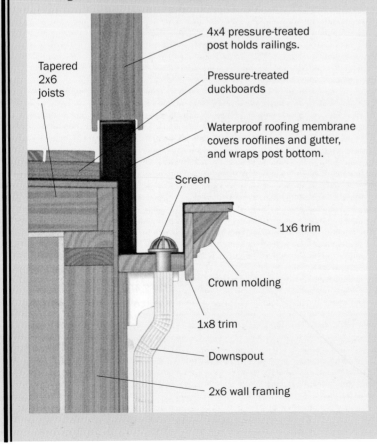

Tapered 2x6 joists

4x4 pressure-treated post holds railings.

Pressure-treated duckboards

Waterproof roofing membrane covers rooflines and gutter, and wraps post bottom.

Screen

1x6 trim

Crown molding

1x8 trim

Downspout

2x6 wall framing

The enlarged, open living room would include a dining area and expand the garden view. It seemed simple.

A Half-Story Keeps a Low Profile

Things that seem simple don't always turn out that way, though. The house was close to the setbacks on two sides of the property, and the other two sides were given over to gardens. This limited how much we could expand the first floor (see the floor plans on p. 40). Another problem was the second story. Because the house is on a hill, a full-height second floor would tower over the property and overwhelm the garden (see the drawings on p. 40). This is precisely what we wanted to avoid: gaining space at the cost of erasing the cottage's unpretentious charm. It was simply too high a price.

The solution was a smaller second story. We raised the outside walls just 3½ ft. from their original height. As a result, the second floor has a kneewall and sloped ceilings, but a shed dormer along the rear of the house helps compensate for the lost space (see the photo on p. 41). The long dormer can't be seen from the front; from the footpath that leads to the front door, the roof profile remains unobtrusively low.

Another part of the solution lay in the large, hip-roofed dormer over the front door. This big dormer creates an open space, or "clerestory," over the front door and stairway (see the photo on the facing page). At the top of the stairway, an open hallway, or "bridge," connects the two bedrooms (the bathroom is in between). This big dormer brought some much-needed light into the staircase and to the front entryway, and it opened the views of the surrounding garden. This summer house has no central air-conditioning, so it was important to keep the second floor as breezy as possible.

Below the first-floor addition, we built a new concrete-block basement with plenty of perimeter drainage. This would be the space for a new gas-fired boiler, which would replace the outdated and expensive electric heat. It would also gain some room for storage. In the original section of the house, we reinforced the first-floor joists and subfloor. We shortened a particularly long beam span with a couple of concrete piers, and we reinforced the joists by sistering new 2x6s onto the existing joists.

Garden Views Are Everywhere

Creating garden views was something we did everywhere we could. Windows surround the great room, which takes up most of the first floor (see the top photo on p. 45). There are so many windows and such dense foliage in the garden that in spring, the living-room walls seem papered with leaves. On the staircase, a circular window offers another view. Upstairs, there is a small deck outside the master bedroom overlooking the garden, enough room for two chairs and a small table (see the photo on p. 44). When the weather is warm, the separation between indoors and out is hardly noticeable.

This connection extends to the front porch, which is a more generous space than in the original cottage. Porches are transitional spaces, part interior and part exterior. We tried to push this concept by adding the hip-roofed, templelike space for an outdoor

Natural light warms the front entry. The hip-roofed dormer over the front door illuminates the front entry and upstairs hallway. From the upstairs, the big window affords a sweeping view of the garden. Photo taken at D on the floor plan (p. 40).

The master bedroom gets a terrace. A small deck over the dining area is just large enough for two chairs and a table. Photo taken at E on the floor plan (p. 40).

dining table and chairs (see the bottom photo on the facing page). This area has the proportions and detailing of an interior space; it's almost an extension of the living room, even when viewed from inside. The new porch itself is deeper than the original, and the detailing is more formal, making it feel like an indoor space even though it is surrounded by the open air.

Deep porches pose a problem—the deeper and more spacious the porch, the darker the adjacent interior space. This is where the clerestory came in handy once again. The amount of light that pours through the big dormer window counteracts the dark shadows cast by the porch (see the photo on p. 43).

Harry N. Pharr has 30 years of experience in the fields of construction and architecture. His experience includes project management, planning, design, architectural documentation, and construction administration for a wide range of projects. The builder was Joe Nachtigal of Nightingale Construction in Warwick, New York.

With foliage like this, you don't need wallpaper. In the combined living and dining room, large windows and a bay-shaped bump-out provide sweeping views of the garden and porch. Photo taken at C on the floor plan (p. 40).

A porch is just an outdoor room. The deep, formal front porch takes full advantage of the spectacular garden view. The hip-roofed, templelike dining area has the proportion and scale of an interior room, without the windows. Photo taken at F on the floor plan (p. 40).

Jewelbox Bathroom

■ BY JEFF MORSE

As an architect who designs a fair amount of remodels, I believe that any changes made to a house should respect or rediscover its original style and individuality. Unfortunately, many of the houses here in Petaluma, California, where I work, have been added to poorly, becoming chaotic jumbles of styles and spaces.

Such was the case with the Rathkey place. John and Cynthia Rathkey own a late nineteenth-century bungalow near the center of town. Their house had been defaced by a flat-roofed 1950s addition off the kitchen and a ramshackle bathroom that was tacked to the side-yard wall. They called me for advice about healing the house—taking away the grotesque additions and adding a garage that would include a master bedroom and a bath above it.

I told the Rathkeys that we could most likely squeeze their addition into the slightly trapezoidal space alongside the house. But keeping the two-story addition in scale with the original bungalow was a problem I immediately began to struggle with: A conventional two-story addition would dwarf the house.

Fortunately, the house had a wood-framed floor over a crawl space. Because the garage floor would be a slab-on-grade, its top surface would end up 30 in. lower than the floor of the house. To scale down further the two-story bulk of the addition, I proposed a story-and-a-half design with the top plate of the upperstory wall set 5 ft. 6 in. off the floor. A 7-ft. 6-in. garage ceiling would bring the top plate of the second-story wall within 2 ft. of the top plate of the original building.

Breaking the Plate

Visually relating the addition to the house was as simple as using the same hip-roof form, with its unusual 7½-in-12 roof pitch. The low plate would bring the ceiling planes much closer to the floor than is typical, creating a dynamic and sheltering space. And to emphasize that volume, I designed the roof to transfer its loads without the need for posts or collar ties that would detract from the space. The low plate, however, created a problem with the windows—especially in the bathroom. Lifted as high as possible, the upper glass line was still well

Squares, triangles and curves. A tub raised atop a low platform occupies its own niche in the bathroom, where it is framed by a complex arrangement of roof lines, archways and wingwalls. Note how the ceiling over the tub is recessed for showering headroom. By carefully laying out different size tiles that share the same module, the author was able to use discounted tiles to create rich geometric compositions.

Directing the Tension

To get the standing eye-level windows in the bathroom dormer, the top plate of the wall had to be interrupted. The detail at right shows how the tension loads from the roof are transferred by steel ties, headers, and beams around the window opening.

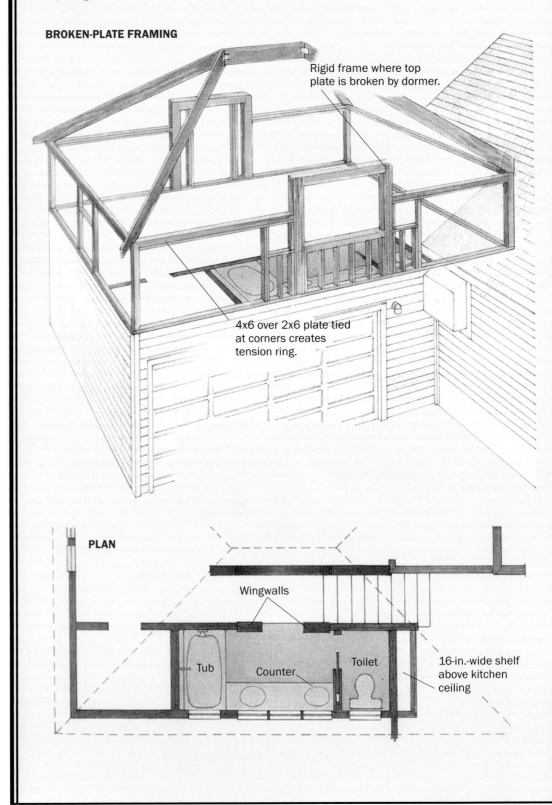

BROKEN-PLATE FRAMING

Rigid frame where top plate is broken by dormer.

4x6 over 2x6 plate tied at corners creates tension ring.

PLAN

Wingwalls

Tub

Counter

Toilet

16-in.-wide shelf above kitchen ceiling

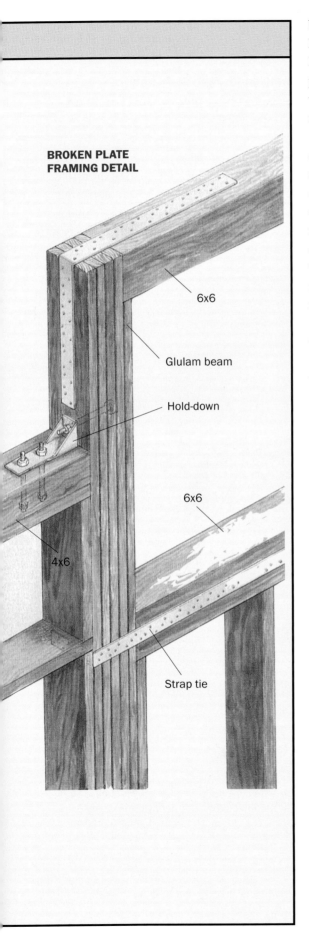

**BROKEN PLATE
FRAMING DETAIL**

6x6

Glulam beam

Hold-down

6x6

4x6

Strap tie

below standing eye level. As important as the low plate was in scaling down the addition, I felt it was equally important to break the plate somewhere to get a standing eye-level view and avoid an oppressive feel to the low roof near the walls.

A shallow-hipped dormer was the perfect device to open up the bathroom and scale down the somewhat monolithic form of the addition's exterior, but it presented an interesting structural dilemma. Because there are no posts to hold up the ridge beam and no collar ties to keep the walls from spreading, the hip rafters take on the load in compression while the top plates act as a tension ring to contain the thrust of the hips.

It doesn't take a genius to figure out what would happen if you broke the plate of such a roof, but it did help to have one on hand when I decided to try it. Richard Hartwell, an engineer and a good friend of mine, devised a scheme to transfer the tension load in the top plate around the window and door openings at the north and south dormers. By installing small glulam columns at either side of the opening, we were able to transfer the tension load in the wall plate to the header and the sill plate at the opening, then back to the opposing wall plate (see the top drawing on the facing page). Simpson® HD2A hold-downs transfer loads from the double plate to the columns, and framing straps transfer it from the column ends into the header and sill plates (see the drawing at left).

An Open Plan

Keeping an open and airy feel to the upstairs addition was something we all wanted, so the Rathkeys readily agreed to opening the bath and the vanity to the main bedroom space (see the photo on p. 52). This also allowed the bathroom to feel expansive well beyond its rather modest dimensions. The bathroom is tucked into a space that is merely 5 ft. wide and 13 ft. long (see the bottom drawing on

Visually relating the addition to the house was as simple as using the same hip-roof form.

p. 48), and about half of that space is taken up by the counter and the bathtub. The arched entry to the bath repeats the shape of the bed's headboard, creating a rhythm of arcs at the top of the stairs.

Exotic Windows

I thought it was important to have some good-size operable windows in the bathroom to balance the light and provide some cross ventilation. Unfortunately, large bathroom windows facing the street are usually ill-advised. Another strike against the windows was that the vanity would be right in front of them. My first thought was to work mirrors into the glazing, but the backside of a mirror isn't exactly a handsome exterior detail.

Then it occurred to me that iridized glass would probably work well. Iridized glass has a thin metallic coating on it that gives it a shimmery, rainbow look resembling a film of oil on water (see the photo on the facing page). This bathroom has two kinds of glass with the iridized coating—

Flush-framed dormer. The author prevents the addition from overshadowing the original house by keeping the second-floor walls a mere 5 ft. 6 in. high at the top plate. A dormer over the vanity provides headroom and space for a trio of windows, which are glazed with iridized glass (detail on the facing page).

water glass and opal glass (Spectrum®
Glass, P. O. Box 646, Woodinville, WA
98072; 425-483-6699).

Water glass is clear with a wavy pattern.
But once the glass is iridized, its reflective
quality renders it completely obscure to
vision from the outside during the day
when the light is brighter outside than in,
and vice versa. Opal glass, on the other
hand, has milky, opaque white swirls in it
that abstract images from either side, no
matter what the lighting.

There are three windows over the lavatory.
Two of them are centered over the sinks and
glazed with mirrors on the inside and
iridized water glass on the outside. The window in the middle also has iridized water
glass on the outside, but the interior side is
glazed with Spectrum's "hammered" glass—
an obscure glass that has a rippled surface.
Each window has nine lights: one large pane
in the middle bordered by narrow strips of
iridized water glass and iridized opal glass.

The final piece of the bathroom-window
puzzle was the need for tempered glass in
the tub area. Much as I tried, I couldn't find
anyone who would temper iridized glass. I
was told that it simply couldn't be done. So
we decided to use glue-chip glass for the
center light of the windows over the tub and
the toilet. Glue-chip glass has patterns
etched into its surface that look like frost.
Glass artisans make the patterns by applying
hide glue to a piece of glass. When the glue
dries, it peels thin chips of glass away in
complex, feathery patterns that let in light
while obscuring the image on the other
side. Best yet for our purposes, glue-chip
glass can be tempered after the patterns are
created in it.

Specialty glass isn't cheap. The iridized
glass cost about $9 per sq. ft.; hammered
glass cost $5.50 per sq. ft.; and the tempered
glue-chip glass came in at $9.50 per sq. ft.
But used judiciously in a prominent place, a

Glazing Detail

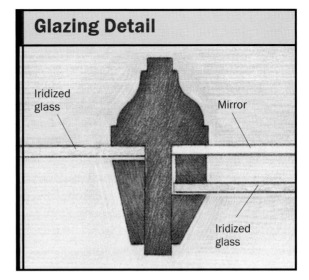

Iridized glass

Mirror

Iridized glass

spot of expensive glass can easily carry its
budgetary burden.

Like the other windows in the addition,
the perimeter lights in the tub and toilet
enclosures are just a fraction under 3 in.
wide. This allowed us to use iridized glass
because the code accepts nontempered glass
3 in. in width or less near tub enclosures.

So who would we get to make such
bizarre windows? We decided early in the
design to stay with wood sash and jambs
(in spite of the additional cost) because they
are an important element in defining the
character of the project, and they're a link
with the original bungalow. I got bids from
three window companies: two large manufacturers and a custom-window builder that
I had been using on other projects. To my
surprise and delight, Dave Ferguson of
B&D Window and Sash Co. (B&D Window,
12590 Oak St., Clearlake Oaks, CA 95423;
707-998-3717) not only could provide the
true divided-light sash at a price equal to
the least expensive production windows but
also could do it in clear heart redwood. Plus
he was able to handle our strange glazing
requirements. Ferguson built the nine-light
windows so that the center light was double
glazed, and the edge lights were single
glazed (see the drawing above). This allowed

Bath alcove. An archway inspired by the headboard in the foreground frames the bath, which includes a separate enclosure for the toilet. The mirrors over the sinks are the interior panels of dual-glazed windows.

a reasonably narrow muntin size, which matches the other windows. For one final wrinkle in the window saga, the mirrored lights over the sinks extend all the way to the top and the bottom of the sash.

Toilet and Tub

To complete the bathroom-as-alcove idea, the toilet was given its own small room (3 ft. by 5 ft.) with a door. The wall of the toilet enclosure on the side of the original house fortuitously abuts the old attic. The attic

amounted to unused space, so we carved a 16-in.-deep linen storage area into it that is tall enough to fit three shelves.

The tub was set at counter height to provide an easy view out the window. But because of the low plate height, raising the tub was accomplished only by extending the ceiling upward into the 12-in.-deep rafter recess, giving adequate headroom to comfortably use the shower. Contractor Larry Volat doubled the rafters on either side of the headroom recess and put 2x4 drywall backing on edge between them to keep the ceiling as thin as possible.

The tub we settled on, the Tea for Two by Kohler® (Kohler Co., 444 Highland Ave., Kohler, WI 53044; 414-457-4441), has a nice combination of ingredients. It was the smallest (32 in. by 60 in.) double-ender I could find, and it comes with or without whirlpool jets. It's the perfect tub for conservation-minded, hedonist couples with space limitations.

Laying Out the Tile

Whenever I look at tilework, the first things I notice are the mistakes—the "Oh, didn't quite fit" cuts. Tile can be pretty unforgiving. It can also be expensive and way too plain in shape, color, texture, and finish for my taste. These factors all influenced the approach we took to the tilework in the Rathkey bathroom, which was executed with care and precision by tile setter Eugene Dolcini.

Tile maker Bob McIntyre runs a modestly sized porcelain and stoneware tile manufacturing operation in a couple of warehouses just off the railroad tracks in Healdsburg, 40 miles north of Petaluma (McIntyre Tile Inc., 55 West Grant St., P.O. Box 14, Healdsburg, CA 95448; 707-433-8866). He produces some of the most beautiful tiles I have seen. What I like best, however, are his seconds. These are tiles that are rejected for a variety of reasons: too wide a color range for the given glaze, iron spotting, or perhaps some slight warping. To me, most of these seeming imperfections add interest and character. The cost of the seconds is a quarter to a third of the firsts, which is also appealing. The trick to using them is being able to cope with the usually limited quantities of seconds available in any given size with a particular glaze.

To that end I've taken to mixing sizes within a field. Since many of McIntyre's tiles are modular, such as 3 in. by 3 in., 3 in. by 6 in., and 6 in. by 6 in., they can be combined into larger patterns to create a random ashlar effect. We used all three sizes on the blue floor (see the photo on p. 47). On the shower wall and tub surround, the green tiles are 3 in. by 3 in. and 6 in. by 6 in. We had enough of the smaller tiles to use them as borders, thereby outlining the field tiles and avoiding tapered cuts in the larger ones. Where the 6-in. by 6-in. tiles abut one another, they are always offset by the width of at least one of the smaller tiles to avoid a static, heavy spot of tile in the field. Similarly, the light and dark tiles are distributed around the wall to keep distracting dark or light patches from developing. The only drawback I have found to mixing sizes in the field using seconds is that they vary slightly in thickness and therefore take more care (and more time) to set.

Once we had the tile and the layout in mind, we measured for trim pieces, such as surface bullnose and quarter-rounds. One of the nice things about working with a custom tile maker like McIntyre is the possibility of special orders. For example, the step into the Rathkeys' tub has an S curve to its front, allowing one to stand in front of the lavatory near the tub. I sent a template to McIntyre's production foreman, and he shaped the appropriate quarter-rounds so that we could achieve a smooth curve without cutting the trim into small pieces. The rounded edge of the bullnose is easy on the feet while getting in the tub, and the step makes an equally good place to sit after a soak.

* Price estimates noted are from 1992.

__Jeff Morse__ is an architect based in Petaluma, California.

Adding a Second Story

■ BY TONY SIMMONDS

It was a wonderful, prematurely warm day at the beginning of March 1994 when I first met Paul and Letizia Myers to discuss adding a second story to their house (see the top photo on the facing page). Both of their children were in their teens, and the house was beyond feeling cramped. A second story would give Paul and Letizia a master suite, a room for each child, and another bathroom.

That sunny March day had the kind of morning when tearing off the roof seems like the most natural and logical thing in the world. In fact, as Paul and I stood in the warm sun and looked at the roof he had repeatedly patched with elastomeric compounds, it seemed an unreasonable strain on anybody's patience to formulate a program, draw plans, and apply for permits.

In reality, the timing should have been perfect. The design could get done, and the plans drawn, in time to begin construction by late summer. August and September are the most reliably dry time of year in Vancouver.

But events foiled us. A strike at city hall slowed the permitting process, and it was into November by the time we had approval to go ahead. Reluctantly, we shelved the project until spring. Then I met contractor Walter Ilg. Walter makes a specialty of han-

dling what he calls "the hard parts" of any renovation. I watched his crew remove and replace the foundation of my neighbor's house, and I was impressed with the expeditious way he handled the hard part of that one. So I showed him the plans for the Myerses' project. We agreed that the way to do it was to put up the new roof before taking down the old one. But we disagreed about timing. I had in mind the end of April. "Why wait?" he said. "It can rain anytime here."

It could do more than that, as were to find out. But on a warm Monday in March, almost exactly a year after my first visit with the Myerses, Walter and his crew started building scaffolding.

It's just too small. Charming in its simplicity and located in a good part of town, this one-story house had been outgrown by its owners. Adding a second story solved the space problem, and using simple construction methods, including prefabricated trusses, kept the total cost to just over $100,000 (Canadian).

Prefab Trusses and Minimal Walls Go up Quickly

Walter's theory of framing is simple. You do the minimum necessary to get the roof on, throw a party and then back-frame the rest. In this case the minimum was less than it might have been because the existing attic floor framing—2x8s on 16-in. centers—didn't have to be reinforced. Not that the job couldn't have been done the same way even if the existing joists had needed upgrading.

The new roof was also designed with minimums in mind: minimum cost and minimum delay. There would be no stick-framing; instead, factory-supplied trusses would carry the loads down the outside walls. Almost half of the trusses would be scissor trusses for the exposed wood ceiling over the stairs and in the master bedroom.

The 12-in-12 pitch apron that forms the overhang at the gables and at the ground-floor eaves would be framed after the new roof was on and the exterior walls built.

To get the roof on, we needed just two bearing walls. But a continuous wall plate couldn't be installed without severing the old roof from its bearing. The solution was to use posts and beams, and to frame in the walls afterward.

Based on the layout of the interior walls, Walter and I decided to use four 4x4 posts along each side of the house. The beams would be doubled 2x10s. In one place, one beam would have to span almost 16 ft., but any deflection could easily be taken out when the permanent wall was framed underneath it. As it turned out, there wasn't any deflection.

So on Tuesday morning, with the scaffolding built, Walter's crew cut four pockets in the appropriate locations along each side

Preparing for the new roof. The crew begins construction of the new roof by excavating post holes in the old roof over the wall plate. On the left, a ramp for removing roof debris leads to a curbside Dumpster.

Posts carry a wall beam. Well-braced with diagonal 2x4s, 4x4 posts rise from the holes in the roof to support a doubled 2x10 beam. Note the temporary flashings that are at the base of the posts. At the far end, the wall beam extends beyond the plane of the house to create a staging area for the roof trusses.

of the roof (see the photo on p. 56). Then they secured the 4x4 posts to the existing floor framing and to the top plate of the wall below. They notched the end posts to fit into the corner made by the end joist and the rim joist. We were lucky with the intermediate ones; all of them could be fastened directly to a joist, notching the bottom of the 4x4 as required. None of the four intermediate locations was so critical, though, that the post couldn't have been moved a few inches in one direction or the other if necessary.

Walter used a builder's level to establish the height of the posts, and by Tuesday afternoon one of the beams was up and braced back to the existing roof (see the photo above), and the posts were in place for the other one.

At the same time, the rest of the crew was cutting away the ridge of the existing roof to allow the flat bottom chord of the common trusses to pass across (see the photo on pp. 54–55). They were able to leave the old attic collar ties/ceiling joists in place, though, because the old ceiling had been only 7 ft. 6 in. I stopped by at the end of the day to inspect the temporary post flashings the crew had made with poly and duct tape. It had been another sunny day. By afternoon, however, thin clouds had moved in, and it was getting cold. The forecast was for snow.

Snow and Rain Complicated the Job

The order for the trusses had been placed the previous week, with delivery scheduled for Thursday or Friday. But on Monday, while we were overseeing the lumber delivery,

The old roof comes down. With the new roof in place, the old one can come down. Next, the missing studs in the perimeter walls will be installed.

Walter let me know that he had called the truss company and promised them a case of beer if they delivered the trusses on Thursday and three cases if they got them here by Wednesday.

On Wednesday morning there was 8 in. of snow on the ground—and on the Myerses'

roof. But the weather system had blown right through, and by 8 AM the snow was melting fast. Walter called to say he had sent two men to sweep the snow off the roof and that the trusses would be on site if the truck could make it out of the yard. At noon I arrived to see the last bundle of trusses being landed on temporary outrigger beams.

The rest of that day was spent finishing the beam, and setting and bracing the trusses. Plywood laid across where the old ridge had been scalped made it easy for one man to walk down the roof supporting the center of the truss while two others walked it along the scaffolding.

Even though it violated Walter's get-it-done roofing rule, I had the crew install the frieze blocking as the trusses were installed. By cutting the blocks with a chopsaw, you can ensure perfect spacing (even layout becomes unnecessary where framing proceeds on regular centers), and it's much easier to fasten the blocking this way than it is to go back and toe-nail it all afterward. Also, the soffit-venting detail I used with the exposed rafter tails required the screen to be

A Prefabricated Roof Apron

The horizontal roof apron that runs along the front and back of the house was assembled with 8-ft.-long, prepainted sections of rafters made up in the shop.

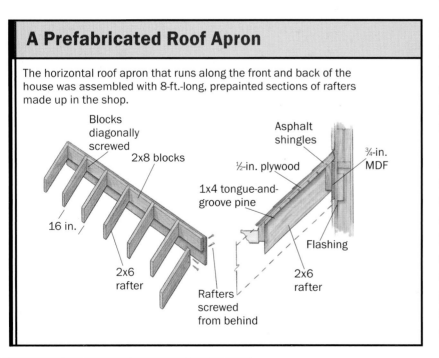

Blocks diagonally screwed

2x8 blocks

16 in.

2x6 rafter

Rafters screwed from behind

Asphalt shingles

½-in. plywood

1x4 tongue-and-groove pine

¾-in. MDF

Flashing

2x6 rafter

sandwiched between two courses of soffit and stapled to the inside of the frieze block.

On Thursday another front brought wind and rain, which dispatched the last vestiges of snow but made a miserable day for the sheathing crew. Having to install the soffit, the screening, and the 2x4 purlins that tie all of the trusses together didn't speed things up. Nor did the four skylights. I didn't want the bargeboards done hurriedly, so to make things easier for the roofers, we temporarily toe-nailed 2x4s on the flat to the trimmed ends of the rake soffits. That way, the roofers could cut their shingles flush to the outside edge of the 2x4, and when the 2x4s were removed and replaced by the permanent 2x10 bargeboard and 1x3 crown, the shingles would overhang by a consistent 1¼-in. margin.

On Friday morning the roofers went to work on one side of the roof while the last nails were pounded into the sheathing on the other side. It didn't take long for them to lay the 12 squares we needed to make everything waterproof. Meanwhile, Walter and his crew were removing the old roof underneath (see the photo on the facing page) and carrying it to the Dumpster in 4-ft. by 12-ft. chunks. I usually try to save the old rafters, but in this case I'm afraid I let the momentum of the job dictate the recycling policy.

By 1 PM, true to his word and to long European tradition, Walter was tying an evergreen branch to the ridge, and plates of cheese, bread, and sausage were being laid out on a sheet of plywood set up on sawhorses in the 26-ft. by 34-ft. pavilion

A roof apron recalls the original proportions. Strong diagonal lines drawn by the 12-in-12 rake boards at the gable ends help to break up what would otherwise be a top-heavy facade. The lower roof continues across the front and back of the house, sheltering the windows and preserving the original roofline.

Bump-out and fascias support the rake soffit. At the gable end, a bump-out protects the upstairs windows and supports the tops of the 2x10 bargeboards. The 2x10s are borne by 2x6 fascias cantilevered past the roof-apron rafters. Note how a built-up water table makes a clean line between the old stucco and the new.

Daylight in the center of the house. Skylights over the centrally located hallway light up the stairs, as well as the bathroom, by way of its generous transom. Photo taken at A on the floor plan (below).

that now occupied the top floor of the house. It might be a little breezy, as Paul said to me over a glass of wine, but at least it was dry.

A Roof Apron Prevents a Boxy Look

It took another three weeks to complete the framing and to do all of the picky work that's an inevitable part of tying everything together in a renovation. One detail, and an important element of the design, is the roof apron that encircles the house to break up the height of the building (see the left photo on p. 59). The apron forms eaves along the front and back of the house. At the gable ends, the apron becomes a rake that rises to the peak of the roof, drawing long diagonal lines across what would otherwise be a tall, blank facade. The effect is of a 12-in-12 roof with 4½-in-12 shed dormers.

The apron has practical value, too, particularly at the eaves, where it covers the top edge of the existing wall finish, providing an overhang to protect the ground-floor win-

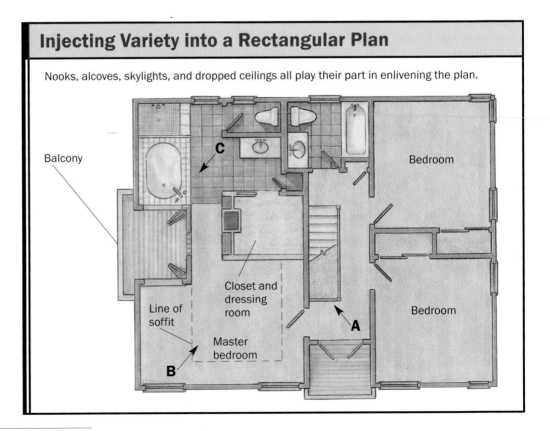

Injecting Variety into a Rectangular Plan

Nooks, alcoves, skylights, and dropped ceilings all play their part in enlivening the plan.

Balcony

C

Bedroom

Closet and dressing room

Line of soffit

A

Bedroom

Master bedroom

B

dows. If you're building outside the painting season, it's essential to get a coat of paint on everything before it's applied to the outside of the house, so we built as much as we could of this apron in 8-ft. sections in my shop (see the drawing on p. 58). For example, the eaves consist of 2-ft.-long 2x6 lookout rafters screwed from the back to a 12-in.-wide strip of Medex® (Medite Corp., P. O. Box 4040, Medford, OR 97501; 800-676-3339), an exterior-grade medium-density fiberboard (MDF) that is gaining popularity for use as exterior trim here. Frieze blocks cut from 2x8s act as pressure blocks between the rafters. We prepainted these assemblies and the 1x4 tongue-and-groove pine that we nailed to their tops in our shop.

On site, the eaves sections were installed and tied together with the prepainted 1x4s and 2x6 fascia. Then we snapped lines on the gable ends from the ridge to the eaves lookouts to establish the line of the rake soffit (see the right photo on p. 59). On this line, we toe-nailed a triangular bump-out, framed out of 2x10s, to the gable-wall framing. From the base of the bump-out, we ran a 2x6 that acts as a rake trim board for most of its length and then becomes the last lookout rafter where it runs into the eaves overhang.

We nailed preassembled and prepainted strips of soffit to the rake trim and to the gable bump-out. Made of tongue-and-groove 1x4s blind-nailed to 18-in.-wide strips of ½-in. plywood, the 8-ft.-long strips of rake soffit were pretty floppy until the 2x10 bargeboards went on.

Projecting the gable peaks out from the plane of the wall did more than provide solid support for the rake apron with its heavy bargeboard. It also created some visual interest and gave a little protection to the bedroom windows in the east wall. The peaks were finished with louvered vents and 1x4 bevel siding. These peaks make a nice big triangle of painted woodwork to balance the large areas of stucco.

A balcony separates the bedroom and the bath. Along the west wall, three alcoves with low ceilings have distinct functions. In the foreground, the shower and tub occupy the first alcove. In the middle, a small balcony overlooks the secluded backyard. In the distance, the third alcove provides space for the bedroom dresser. Photo taken at C on the floor plan (facing page).

The outdoors is nearby. On the left, folding windows lead to a balcony off the master suite. On the right, a 7-ft. closet wall separates bedroom from lavatory. The sloping ceiling extends beyond the ridge to become part of the skylight well over the closet. Photo taken at B on the floor plan (p. 60).

We also ran a water table at the second-floor joist level (see the right photo on p. 59). Besides its aesthetic contribution, this band covers the flashing protecting the top edge of the old stucco and makes a practical separation so that new stucco and old don't have to meet. Detailing woodwork so that

stucco always has a place to stop and so that no one panel of it is too big makes the plasterer's job a whole lot easier.

Shaun Friedrich, who learned the stucco trade from his father and who can tell without leaving his truck what a particular stucco is, when it was done, and quite often who did it, made a beautiful job of approximating the look of the original dry-dash finish. Dry dash is a labor-intensive stucco finish in which a layer of small, sharp stones is embedded in a layer of mortar. Shaun rendered a compatible finish for the upstairs walls by using a drywall-texturing gun to create the random, splattered look of dry dash. This substitution saved us $1,000.

Allocating the New Space

On the inside, Walter's crew was turning the 26-ft. by 34-ft. pavilion into a second floor with three bedrooms and two baths (see the floor plan on p. 60). The west end contains a master bedroom and bath. In the center of the house, a hallway includes the existing stair, a bathroom at the north end (see the photo on p. 60) and a balcony at the south end. Bedrooms at the east end complete the plan.

The subdivision of the master-bedroom space to accommodate a walk-in closet and the bathroom was the most intriguing part of the design. I wanted the room to feel large and generously proportioned, but at the same time I wanted the different areas within it to be well defined.

The first division is between north and south. The bathroom, with its requirement for privacy, is on the north side; the bedroom is on the south. What separates them is not a wall but other subsidiary spaces: the walk-in closet and a small balcony (see the photo on p. 61).

Then there is the division between the main part of the room and the three 6-ft.-deep alcoves along the west wall. Linked by their common ceiling height—7 ft.—the alcoves contain from north to south the shower/tub space; the balcony; and the bedroom-dresser area.

In addition to their common ceiling heights, the alcoves are further linked by large windows that open onto the balcony (see the photo on the facing page). These windows can be folded back against the wall so that in nice weather the balcony is really a part of the bedroom.

The transparency of these linked alcoves to one another goes a step further. On the bathroom side, the shower is separated from the tub by a glass partition; on the bedroom side a window in the south wall lines up with the two windows to the balcony. Standing in the shower, you can look right through four transparent layers to the outside. In a small house, long views such as these foster a sense of spaciousness.

The ceiling in the master bedroom is an example of how to turn a technical problem to practical advantage. The decision to use trusses throughout for the sake of expeditiousness and economy meant that the ceiling could slope at a pitch of only 2-in-12 (the bottom chord of a 4½-in-12 scissor truss), and that skylight wells would necessarily be rather deep. Locating the skylight so that the slope of one side of the ceiling extends into the skylight well to meet the top of the skylight makes for a dramatic light shaft that spills light all along the ceiling as well as down the wall. It also didn't leave much room for error in the layout we had to do back on that raw day in March when Walter's crew members were swarming over the roof with snow in their hair and shinglers at their heels.

As for the low slope of the ceiling, we made it seem higher by holding the closet walls to a height of 7 ft. In the end, the effect was everything we had hoped it would be. Letizia, who is Swiss and for whom I was trying to echo a wooden chalet ceiling, was not disappointed.

Price estimates noted are from 1996.

Tony Simmonds is a designer and builder in Vancouver, British Columbia, Canada.

The ceiling in the master bedroom is an example of how to turn a technical problem to practical advantage.

A Gable-Dormer Retrofit

■ BY SCOTT MCBRIDE

Carol and Scott Little's home draws its inspiration from the cottages of colonial Williamsburg and the one-and-a-half story homes of Cape Cod. Both styles typically feature a pair of front-facing gable dormers. But for some reason, the builder of the Littles's house put only one dormer on the front, leaving the facade looking unbalanced. I was hired to add a new dormer on the front of the house to match the existing one and, for more light, added a scaled-down version of the same dormer on the back of the single-story wing.

As the crew set up the scaffolding and rigged the tarps against the possibility of rain (see "Protecting the Roof" on p. 66), I crawled under the eaves to study the existing roof. I soon realized that framing the sidewalls of the two dormers and directing their load paths would require different strategies, as would the way the dormer ridges would be tied to the main roof.

The first consideration in a retrofit is the location of the dormers, and the second is their framing. The existing front dormer fit neatly into three bays of the 16-in. on center main-roof rafters. These main-roof rafters (or commons) were doubled up on each side of the dormer, creating the trimmer rafters that

carry the roof load for the dormers. Full-height dormer sidewalls stood just inside these trimmers, extending into the house as far as the bedroom kneewalls. Additional in-fill framing completed the dormer walls that were above the sloped bedroom ceiling.

Fortunately, three rafter bays at the other end of the roof landed within a few inches of balancing with the location of the existing dormer. Consequently, I had only to sister new rafters to the insides of the existing ones to form the new trimmers, and I could match the framing of the existing front dormer, leaving a uniform roof placement, appearance, and size.

Cut the Opening and Shore Up the Framing

After laying out the plan of the front dormer on the subfloor, I used a plumb bob to project its two front corners up to the underside of the roof sheathing. Drilling through the roof at this location established the reference points for removing the shingles and cutting the openings.

The tricky part was establishing how far up the slope to cut the opening. To play it

Protecting the Roof

To protect the exposed roof against rain, we rolled up new poly tarps around 2x4s and mounted them on the roof ridge above each dormer. The tarps were rolled down like window shades each evening, with some additional lumber laid on top as ballast. The ballast boards were tacked together as a crude framework so that they would not blow away individually in high winds.

safe, I first opened just enough room to raise the full-height portion of the sidewalls (see the photo and drawing on the facing page). With those walls up and later with some dormer rafters in place, I could project back to the roof to define the valley and then enlarge the opening accordingly.

Inserting new rafters into an already-sheathed roof can be problematic because of the shape of the rafters. They are much longer along the top edge than along the bottom, so there's no way to slip them up from below. A standard 16-in. bay doesn't afford nearly enough room to angle them

in, either. To form the new trimmer rafters, we cut the new members about 6 in. short of the wall plate before we secured them to the existing rafters.

When faced with this situation, I normally use posts to transfer the load from the trimmer rafters to an above-floor header. In fact, I did follow this step with the smaller rear dormer (see "Back Dormer Demands Different Strategies" on p. 68), but that would not work in this case. Here, the floor joists ran parallel to the front wall, instead of perpendicular to it, and so could not transfer the load to the wall.

Supporting the Dormer

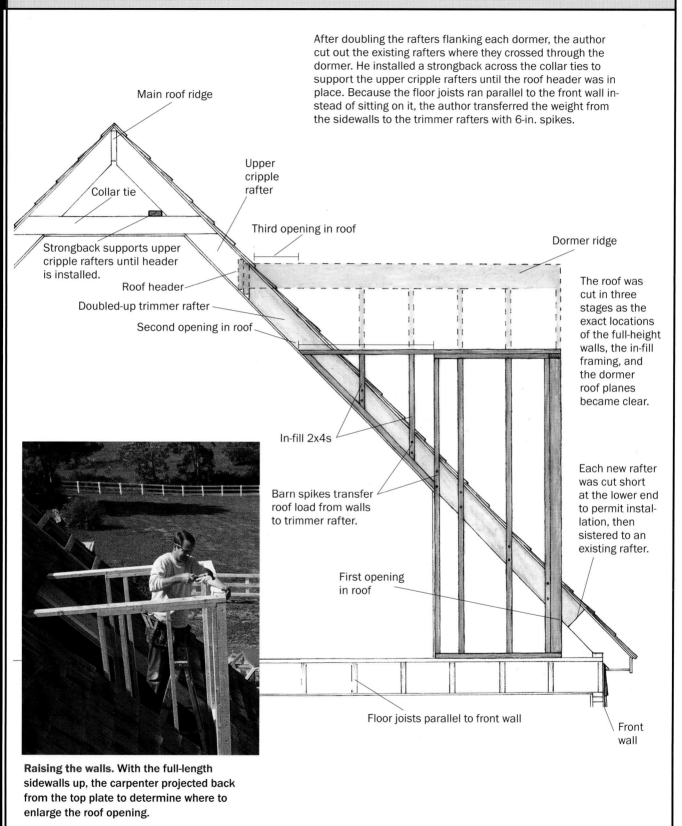

After doubling the rafters flanking each dormer, the author cut out the existing rafters where they crossed through the dormer. He installed a strongback across the collar ties to support the upper cripple rafters until the roof header was in place. Because the floor joists ran parallel to the front wall instead of sitting on it, the author transferred the weight from the sidewalls to the trimmer rafters with 6-in. spikes.

Main roof ridge

Collar tie

Upper cripple rafter

Strongback supports upper cripple rafters until header is installed.

Third opening in roof

Dormer ridge

Roof header

Doubled-up trimmer rafter

Second opening in roof

The roof was cut in three stages as the exact locations of the full-height walls, the in-fill framing, and the dormer roof planes became clear.

In-fill 2x4s

Barn spikes transfer roof load from walls to trimmer rafter.

Each new rafter was cut short at the lower end to permit installation, then sistered to an existing rafter.

First opening in roof

Floor joists parallel to front wall

Front wall

Raising the walls. With the full-length sidewalls up, the carpenter projected back from the top plate to determine where to enlarge the roof opening.

Back Dormer Demands Different Strategies

Unfortunately, when it came to the smaller dormer in back, the existing rafter layout did not match where the dormer needed to be, as was the case in front. Here, I had to build new trimmer rafters in the middle of the existing rafter bays.

The attic space differed, too. Whereas the front dormer served a bedroom, the back dormer was in a storage room. Because the owner wanted to maximize floor space in this storage area, I built the sidewalls on top of the rafters, which pushed the kneewall back and allowed the ceiling slope to extend all the way to the dormer's gable wall (see drawing below).

As in the front dormer, we cut the new trimmer rafters short. This time, however, the floor joists ran perpendicular to the front and back walls, which meant that I could use posts to transfer the load from the trimmer rafters to an above-floor header (see the photo below). The header distributes the weight over several floor joists, and the joists carry the weight back to the wall. The additional strain imposed on the floor joists is minimal because the header is so close to the wall.

The location of the back dormer's roof ridge altered another aspect of the framing. Because this dormer's roof ridge was at the same elevation as the main-roof ridge, I tied the dormer ridge and the main-roof ridge directly together instead of building a separate dormer header.

Measuring and cutting the valley rafters was the same for the back as for the front, with the exception that at the bottom of the back dormer's valley rafters the compound miters did not need a level seat cut because the valley rafters would not sit on top of 2x4 wall plates. Instead, the valley rafters were simply nailed to the face of the trimmer rafters.

ABOVE-FLOOR HEADER DISTRIBUTES WEIGHT OF THE DORMER

There wasn't enough room in the existing roof structure to install full-length trimmer rafters that would bear on the exterior wall. Instead, the trimmers were cut short, and an above-floor header was used to transfer their loads to the floor joists and to the exterior wall.

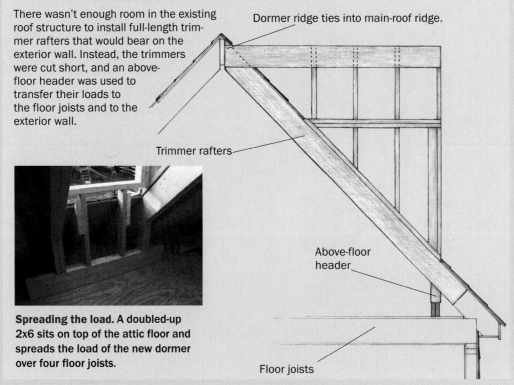

Dormer ridge ties into main-roof ridge.

Trimmer rafters

Above-floor header

Floor joists

Spreading the load. A doubled-up 2x6 sits on top of the attic floor and spreads the load of the new dormer over four floor joists.

I decided simply to let the existing single rafter carry the load for the last 6 in. to the front wall plate. This situation is not the ideal solution, but the weight of the dormer is not great enough to overtax the rafters over such a short span, and doubling up the new trimmer rafters would at least stiffen the existing rafters considerably.

With the new trimmer rafters in, the existing main-roof rafters falling between them were cut and partially torn out to make room for the dormers. The portions above and below the dormer would remain as cripple rafters. The lower cripple rafters were plumb cut in line with the dormer front wall where they would be spiked to cripple studs. Rough cuts were then made at the top, leaving the upper cripple rafters long. These rafters would be trimmed back later, after we established the precise location of the dormer-roof header.

To support the upper cripple rafters temporarily, I climbed up into the little attic above the bedroom. There I laid a 2x4 strongback across four collar ties, including the collar ties connected to the recently doubled trimmers. This strongback would support both the collar ties and the upper cripple rafters until we could install the dormer-roof header.

Dormer Sidewalls Can Be Framed Two Ways

With trimmer rafters installed and cripple rafters secured, I could proceed with the walls. I know two common ways to frame dormer sidewalls: You can stand a full-height wall next to a trimmer rafter, or you can build a triangular sidewall on top of a trimmer rafter, which is how I framed the rear dormer. To match the new front dormer to the existing one, I used full-height studs 16 in. on center only as far in as the kneewall.

This type of dormer sidewall normally delivers the weight of a dormer to the floor.

In new construction, this weight is taken up by doubling the floor joists under these walls. I didn't want to tear out the finished floor, however, so I joined the full-length sidewalls to the trimmer rafters by predrilling and pounding two 6-in. barn spikes through each stud. This transferred the load to the trimmer rafters rather than placing it on the floor framing. I've seen barn spikes withstand tremendous shear loads in agricultural buildings, so I felt confident they could carry this little dormer.

Plan the Cornice Details before Framing the Roof

With the walls up, the roof framing, which is the most complicated, came next. Before cutting any dormer rafters, though, I drew a full-scale cornice section, using the existing dormer as a model. Worrying about trim before there is even a roof may seem like the tail wagging the dog, but it makes sense, especially in a retrofit. The existing dormer featured a pediment above the window. The eaves had neither soffit nor fascia, just a crown molding making the transition from the frieze board to the roof (see the bottom photo on p. 73). That detail reduced the dormer rafter tail to a mere horn that would catch the top of the crown molding. The eaves section drawing helped establish the cuts for the rafter tails and trim details.

Along the rakes, the crown molding was picked up by the roof sheathing, which was beveled and extended out past the gable wall. Using a short piece of molding as a template, I worked out the amount of the overhang and the correct bevel for the edge of the sheathing in the rake-section drawing. Juxtaposing the drawings ensured that the rake crown, the eaves crown, and the level-return crown would all converge crisply at a single point.

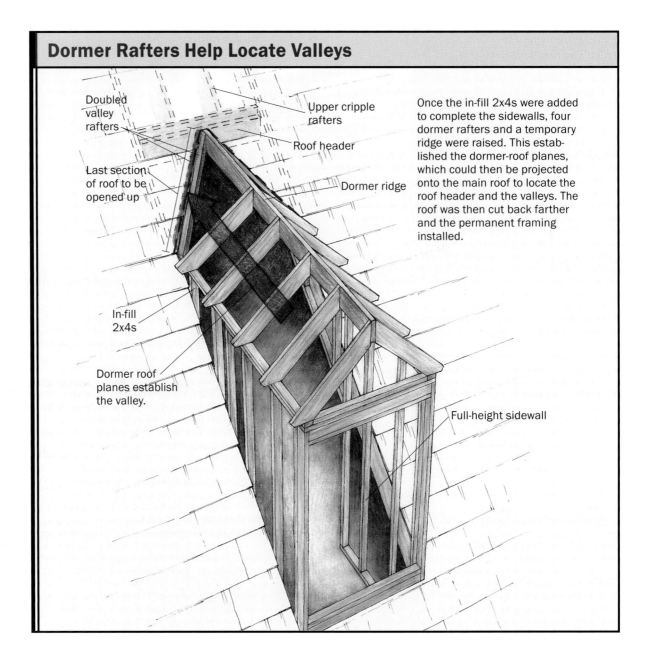

Dormer Rafters Help Locate Valleys

Doubled valley rafters

Upper cripple rafters

Roof header

Last section of roof to be opened up

Dormer ridge

In-fill 2x4s

Dormer roof planes establish the valley.

Full-height sidewall

Once the in-fill 2x4s were added to complete the sidewalls, four dormer rafters and a temporary ridge were raised. This established the dormer-roof planes, which could then be projected onto the main roof to locate the roof header and the valleys. The roof was then cut back farther and the permanent framing installed.

Framing the Roof Defines the Valleys

Ready to proceed with the roof framing, we set up two pairs of common rafters with a temporary ridge board between them. Then we used a straightedge to project the outline of the dormer roof planes onto the main roof and cut back the main-roof sheathing accordingly (see the drawing above). Having established the elevation of the dormer ridge, we trimmed back the upper cripple rafters and then installed the roof header to

carry the permanent dormer ridge board. The roof header spans between the trimmer rafters, carrying the dormer ridge and the valley rafters. (On the rear dormer, the ridge was level with the main-roof ridge, so no header was necessary there.)

When the dormer common rafters and ridge were installed permanently, we used the straightedge again to find the intersection of the dormer roof planes and the inside face of each trimmer rafter (see the photo on the facing page). This point is where the centers of the valley rafters would meet the trimmer rafters. At their tops, the valley rafters would nuzzle into the right angle formed between the dormer ridge and the main-roof header.

I like to "back" my valley rafters, a process of beveling them so that they accept the sheathing of each adjoining roof on its respective plane. Because a cathedral ceiling was to wrap under the valley, I backed the lower edge of the valley as well, giving a nice surface for attaching drywall.

In addition to backing, I double valleys, even when not structurally necessary, because it gives ample bearing for plywood above and drywall below. Doubling valley rafters also simplifies the cheek-cut layout at the top and bottom of the valley because a single compound miter is made on each piece instead of a double compound miter on a single piece.

Because of the dormer's diminutive size, valley jack rafters weren't required. Consequently, with the valleys in place, the framing was complete, and we could dry it in.

Sheathing and Flashing Combat Wind and Water

We sheathed the front of each dormer with a single piece of plywood for maximum shear strength (see the photo on p. 72). With so little wall area next to the windows, I was concerned that the dormer might rack in high winds. The small back dormer was especially worrisome because it had no full-length sidewalls to combat racking, but the single piece of plywood on its front stiffened the whole structure. We extended the roof sheathing past the gable wall and beveled it to receive the rake crown molding.

Flashing work began with an aluminum apron flashing at the bottom of the dormer front wall (see the top center photo on p. 73). The downhill fin of this flashing extends a few inches beyond both sides of the dormer, and its vertical fin was notched

and folded back along the sidewall. Then the first piece of step flashing had its vertical fin folded back along the front wall to protect the corners where the apron flashing had been notched (see the top left photo on p. 73). Step flashings march up along both sides of the dormer, with the uppermost pieces trimmed to fit tightly beneath the dormer roof sheathing. It was tough work weaving step flashings into the existing cedar-shake roof. Hidden nails had to be

Finding the bottom of the valley. A straightedge is laid across the dormer rafters to project the roof surface to the inside of the trimmer rafter.

In the doghouse. Narrow dormers are prone to racking. To stiffen this one, the author sheathed the front wall with a single sheet of plywood.

extracted with a shingle ripper, a tool with a flat, hooked blade. If I had it to do over, I would sever these nails with a reciprocating saw before the dormer sidewalls were framed.

The valley flashing was trimmed flush with the dormer ridge on one side of the roof, and the opposing valley flashing was notched so that it could be bent over the ridge. We protected the point where the valleys converge at the dormer ridge with a small flap of aluminum with its corners bent into the valley. This approach is more reliable than caulk.

The last piece of flashing to go on was the gable water-table flashing (see the top right photo on the facing page). Its front edge turns down over the return crown molding, and its rear corners fold up under the extended roof sheathing to repel wind-driven rain.

Water-table flashing protects window and trim. The crown that forms the bottom of the pediment will go below the flashed water table and miter with an eaves crown (seen poking out past the corner).

Keep the water moving down. An apron flashing seals the front wall with its ends bent around the corners (center photo). Then the lowest step flashings have their vertical fins bent over to cover the notches in the apron (bottom photo).

Finish Trim Improves Weather Performance

The house is just a few years old, but the existing front dormer had suffered extensive decay. In the worst shape were the finger-jointed casings and sill extensions that the original builder had used. To avoid a repeat of this calamity, I used only solid moldings and bought cedar for the trim boards. Everything was primed, especially the ends. To promote air circulation, the ends of corner boards and rake boards were elevated an inch or so above nearby flashings.

We wanted the new cedar shakes on the dormer to blend in with the existing weathered roof. I asked around for a stain recipe, but the only response I got was from an old farmer standing at the lumberyard counter. He insisted that horse manure was the ticket. To my great relief, we hit upon a more savory alternative. We brushed on an undercoat of Minwax® Jacobean, followed by a top coat of oil-based exterior stain in a driftwood-type shade. The undercoat added a nice depth to the gray top coat.

Scott McBride is a contributing editor of Fine Homebuilding *magazine. His most recent book is* Windows and Doors *(The Taunton Press). McBride has been a building contractor since 1974.*

A crowning moment. Three pieces of crown converge at the bottom corners of the pediment. Trim, casings, and sills—primed on every side—resist rot.

Framing an Elegant Dormer

■ BY JOHN SPIER

Some years ago, my wife, Kerri, and I built a small Cape-style house for ourselves on Block Island, Rhode Island, where we live and work. Most small Capes have essentially the same upstairs plan: a central stairwell and a bathroom at the head of the stairs with bedrooms on each side. Dormers provide the headroom to make these upstairs spaces usable.

The most common arrangement is probably a doghouse, or gable, dormer in each of the bedrooms, with a larger dormer on the other side for the bathroom. Another alternative is a shed dormer over all three spaces, but we weren't too keen on that look. Then at some point, one of us found a picture of a dormer that was essentially two doghouse dormers connected by a shed dormer.

This design would give us as much interior space as a shed dormer, and it was a lot nicer looking. Of course, we argued over the choice at great length. Kerri, the artist, insisted on the beauty and complication of this hybrid dormer (see the photo on p. 76), while I, the practical carpenter, thought about how much easier and faster a basic shed dormer would be. I never had a chance of winning that argument.

As our dormer took shape, an old-timer on the island told us that what we were building was called a Nantucket dormer. The name stuck, and we use it to describe the several different variations that we've built since, including the project in this chapter. Ironically, the history experts on Nantucket island disavow any connection to the name, claiming that the design has no historical precedent.

Two Different Strategies for Two Different Interiors

Even though its design seems to be two dormers connected by a third, the Nantucket dormer is actually built as a single structure. The front wall can be a single plane or its center section can be recessed. The project in this article has the center section stepped back, a look that I've come to prefer. As with most dormers, I think Nantucket dormers look better if the walls are set back from the ends and edges of the main roof and from the plane of the walls below.

I frame Nantucket dormers two different ways to produce two distinctively different

The doghouse-dormer wall is plumbed and braced in place.

Hybrid dormer. Doghouse dormers create more room and larger egress for the bedrooms at the ends of this house, while the shed room in between creates a space for a full bath.

> *The key element in supporting a Nantucket-dormer design is that it is point loaded.*

interiors. The difference, roughly speaking, is that one method uses structural rafters and the other uses structural valleys.

Framing the dormers with structural valleys allows the interior partitions to be eliminated, creating one big open room with interesting angular ceiling planes. For the project in this chapter, however, we used the structural-rafter method to create the more common floor plan with two bedrooms and a bath.

The key element in supporting a Nantucket-dormer design is that it is point loaded, either at the bases of the valleys or at the bottoms of the carrying rafters. Those loads need to be carried by appropriate floor

or wall structures below. The same frame that supports the uniform load of a shed dormer might not carry the point loads of a Nantucket dormer. If you have any doubts at all, it's a good idea to have a structural engineer evaluate the support structure.

Doghouse Walls Go Up First

After the main gables of the house are raised and braced, we lay out the locations of the main roof rafters on the top of the main wall plates. We also locate and snap lines for the outside walls of the doghouses and the shed on the second-floor deck.

The first things that we build are the two doghouse gables. We use the same process that we used for building and raising the main gables, only in a smaller scale. Just as with the main gables, the walls are framed, sheathed, housewrapped, and trimmed before they are lifted and braced plumb (see the photo on p. 75).

Next, we turn our attention to the main ridge of the house. Temporary scaffolding or pipe staging is set up down the middle of the house to work from while the ridge is set. We place the ridge boards (in this case, 2x12s) on top of the plates and transfer the rafter layout directly from plate to ridge (see the photo below).

The ridge boards are set in their pockets and held up with temporary posts and a few common rafters, which help to keep them straight and level.

Structural Rafters and Headers Form the Backbone of the Roof

With the ridge in place, the structural rafters adjacent to the doghouse walls are installed. We doubled these rafters using ½-in. plywood spacers in between to create a total thickness of 3½ in., which matches the width of wall plates for interior partitions below the doubled rafters. If you're building larger dormers, a triple rafter or a double LVL can be used.

The four sets of structural rafters split the main roof into three bays. Our next step is to hang doubled 2x12 headers in each of these bays. The tops of the headers are beveled to match the pitch of the rafters they hang from, and steel hangers hold the headers in place.

Laying out the ridge. To get the rafter layout to match precisely, the layout on the plates is transferred directly to the ridge stock.

The two outside headers carry the valley rafters and the ridges of the doghouses. The center header holds the rafters of the shed section. Along with the structural rafters and the dormer walls, these headers form the backbone support for the dormer-roof structure (see the drawing below).

While the headers are being built and installed, other crew members build and raise the front wall of the shed section. Next, we build and sheathe the triangular sidewalls, or cheeks, on the doghouses that support the common rafters for the two end sections. The doghouse ridges are dropped in

next (see the photo on the facing page), and their common rafters are cut and installed.

Lining Up the Roof Planes and Soffits

Until this point, all the framing has been fairly routine. But now we bump into the chief complication in framing the Nantucket dormer, the fact that the shed roof in the middle is at a different pitch from the gable roofs of the doghouse dormers on each side. In this case, the pitch of the main house

Backbone of an Elegant Dormer

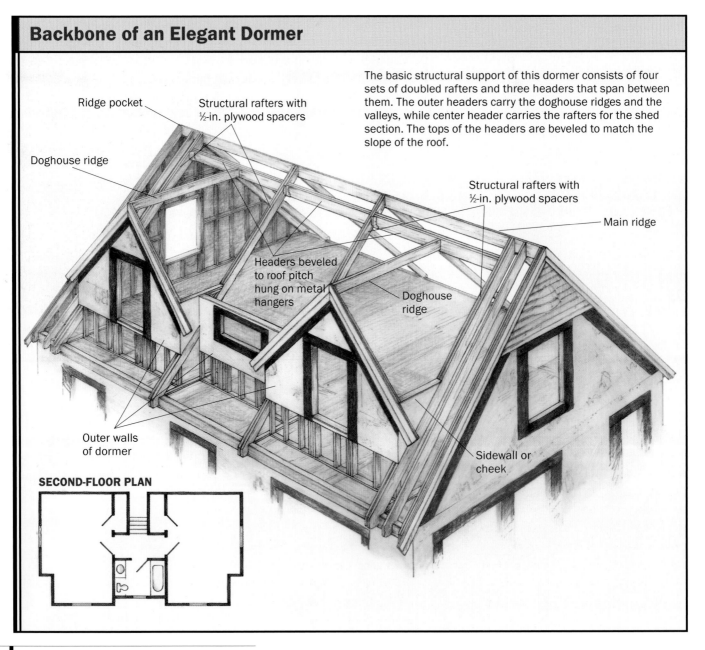

The basic structural support of this dormer consists of four sets of doubled rafters and three headers that span between them. The outer headers carry the doghouse ridges and the valleys, while center header carries the rafters for the shed section. The tops of the headers are beveled to match the slope of the roof.

Ridge pocket

Structural rafters with ½-in. plywood spacers

Doghouse ridge

Structural rafters with ½-in. plywood spacers

Main ridge

Headers beveled to roof pitch hung on metal hangers

Doghouse ridge

Outer walls of dormer

Sidewall or cheek

SECOND-FLOOR PLAN

Doghouse ridge drops in. The doghouse ridge connects the outer wall of the doghouse dormer to the header.

roof and the doghouse roofs was 12-in-12, and the shed roof worked out to be 7-in-12.

Different roof pitches mean that the valleys where the roof pitches intersect are irregular (they don't run at a 45° angle in plan). It also means that the roof planes have to align and that the rafter tails have to be adjusted to get consistent fascia heights and soffit levels. So early on in the process, I work out the rafter details (see the drawing on p. 80). These elements can be worked out on the drawing board, but most often, I make a full-scale drawing of the trim details on either rafter stock or on a sheet of ply-

wood. With this on-site drawing, I can design the rafter tails before I pick up a saw.

Another complication caused by the differing roof pitches is getting the planes of the cathedral ceiling inside to line up. Obviously, a 2x10 rafter meeting a valley at a 12-in-12 pitch will do so at a much different depth than one meeting it at a 7-in-12 pitch. The simplest approach to this problem is to increase the size of the framing material for the shed-roof section. With dormer gables at the same pitch as the main roof, the vertical depth of the rafters at the plate can be measured. The central shed portion of the dormer has a shallower pitch, so

Working Out the Details of the Shed Rafters

The rafters on the doghouse sections are 12-in-12 pitch, and the shed rafters are 7-in-12 pitch. A three-step drawing gets the fascias and soffits to line up, along with the roof and ceiling planes.

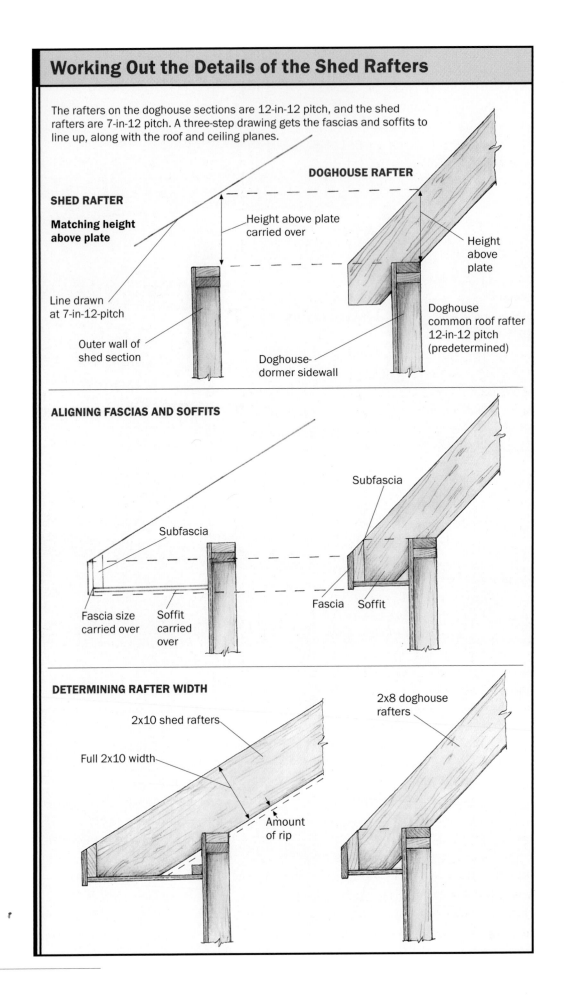

DOGHOUSE RAFTER

SHED RAFTER

Matching height above plate

Height above plate carried over

Height above plate

Line drawn at 7-in-12-pitch

Outer wall of shed section

Doghouse common roof rafter 12-in-12 pitch (predetermined)

Doghouse-dormer sidewall

ALIGNING FASCIAS AND SOFFITS

Subfascia

Subfascia

Fascia size carried over

Soffit carried over

Fascia   Soffit

DETERMINING RAFTER WIDTH

2x8 doghouse rafters

2x10 shed rafters

Full 2x10 width

Amount of rip

it requires a larger rafter size to achieve the same vertical dimension. For this project, the doghouse rafters were made of 2x8s, but the shed-roof rafters had to be 2x10s. But for the roof planes outside and the ceiling planes inside to match up, the 2x10s had to be ripped down to around 9 in. (see "Finding the Exact Rip" at right).

Valleys Are Strung and Measured

Four years of architecture and engineering school taught me that it is possible to work out the framing details of an irregular valley using math and geometry. They even gave me the tools and education to do it. But 20 years as a carpenter have taught me that figuring out irregular valleys is faster and easier with a taut string.

After cutting and installing the common rafters and subfascia for both the doghouses and shed, I stretch a string from the corner where the subfascias meet up to the intersection of the header and the doghouse ridge. From this string, I measure the length of the valley rafters (see the top photo on p. 82) as well as the angles of the top and bottom plumb cuts (see the left center photo on p. 82), the seat-cut angle, and the bevel angles I need to cut into the rafter ends (see the right center photo on p. 82). Armed with all this information, I cut a valley rafter and drop it into place. Then, using a straight-edge from the commons on each side, I mark the bevel cuts on the bottom edge.

After cutting the bevels, I install the valley rafters permanently (see the bottom photo on p. 82) and then lay out and measure the jack rafters. The jack rafters in the center section usually have a compound cut where the miter is beyond the 45° or even 60° that most saws can cut.

If I have a lot of those cuts to make or if the framing is to going be left exposed, I make my cuts with a jig using either a hand-saw or a reciprocating saw to make a clean, accurate cut. However, the typical dormer

Finding the Exact Rip

Once the stock size of the shed rafters has been determined (in this case, 2x10), a simple procedure determines the final rafter width. First, the height-above-plate distance is measured for the 2x8 doghouse-dormer rafters. Next, that distance is transferred to a plumb-cut line at the shed-rafter pitch. That point marks the final width of the rafter.

Height above plate for doghouse rafter at 12-in-12 pitch

Height above plate at 7-in-12 pitch plumb cut

Full width of 2x8   Remove this amount

Final rafter width

Full width of 2x10

Measuring an irregular valley. The quickest way to figure out the irregular valley is to stretch a line from corner to corner. Here, a measurement is taken along that line.

Finding the plumb cut. A rafter square held in the corner against the string determines the angle of the valley plumb cuts.

Corner of the valley. Angles that are taken on each side of the string determine the corner cuts that are needed for the ends of the valley rafter.

Valley rafter slips into place. After all the angles have been cut into the valley rafters, including a bevel on the bottom edge where the ceiling planes intersect, the valleys are nailed in permanently.

has only three or four jack rafters per side, so I mark the angles on the rafter stock and cut them with a circular saw as close as the saw allows. I carve out the remainder of the wood by slowly and carefully dragging the circular-saw blade across the face of the cut. This operation is potentially dangerous, so if you're not comfortable with it, you can use one of the methods mentioned above.

The Rest Is Plywood

While jacks are being cut and fit (see the top photo on the facing page), other crew members fill in the cripple rafters that complete the main roof framing below the three dormer sections. The subfascia is applied to the main eaves, and we can start running the sheathing. Other than the fact that the various plywood shapes are somewhat irregular, sheathing proceeds in the usual fashion, working from the eaves up (see the bottom photo on the facing page).

With the outside framed, sheathed, and ready for roofing, we can turn our attention to the inside. The interior of a Nantucket dormer is usually finished with a cathedral ceiling, which helps the small second-story spaces feel more spacious and airy.

Instead of applying the cathedral-ceiling finishes directly to the bottoms of the rafters, here in New England we nail 1x3 strapping to the rafters, usually 12 in. or 16 in. on center.

Strapping the ceilings of the Nantucket dormer not only simplifies board hanging, providing an extra measure of resistance to deflection and nail popping, but also helps provide a smooth, easy transition between the various ceiling planes. To this end, I usually supplement the strapping by running 1x6 or 1x4 on the undersides of the valley rafters where the roof planes intersect.

John Spier and his wife, Kerri, have a general-contracting and renovation business on Block Island, Rhode Island. His book, Building with Engineered Lumber, *is forthcoming from The Taunton Press.*

Jack rafters complete the framing. With the valleys in place, jacks are cut and installed to finish the framing of the roof planes.

Skinning the dormer. When the framing is complete, the sheathing is applied, bringing out the final dramatic shape of the Nantucket dormer.

Keeping a Dormer Addition Clean and Dry

■ BY NICHOLAS PITZ

Retrofitting a shed dormer in an occupied house can be a disruptive project. But with careful planning, intrusions such as a constant parade of workers and demolition debris can be kept out of the house. And although there's always some time when the roof is open or a wall is missing, avoiding weather damage is straightforward: Just expect pouring rain every night, and plan accordingly.

First, Build the New Roof

The work on this house involved replacing two small existing dormers with a large shed dormer, and the job affected almost every upstairs room. To ensure that the project would be weathertight, we'd build the new dormer's roof first, but we needed room for demolition and reconstruction under it.

We began by framing the dormer's outside wall with four posts and a long header. This approach meant we could frame the dormer roof while cutting only four small holes in the original roof. Then, after we built the new roof and removed the old dormers, we could fill in the studs and the window framing.

We marked the locations of the wall posts to correspond with where we wanted the interior partitions to fall. Then we cut out small sections of roof with a reciprocating saw to give us access to the top plate of the existing wall and installed the first corner post. When this post was carefully braced and plumbed, we used a water level to calculate the height of the other corner post to ensure that the roof was level. Once this second post was positioned and plumbed, we strung a line between the two end posts and set the remaining posts to that height, making sure the load was transferred directly to a stud below.

With the posts braced and in place, we installed the headers. At the end of the day, we covered the holes in the roof made by the new posts with aluminum flashing and plastic sheeting (see the the top drawing on p. 88).

The small dormers had to go. The original house lacked space upstairs, so a big shed dormer was needed. To minimize disruption for the folks living in the house, access to the work zone was via ladders and scaffolds, not through the house.

Tear Out Only What Can Be Rebuilt That Day

We framed the new dormer's roof in two sections over two days. We calculated where the new dormer rafters would intersect with the oldest section of roof, cut a 2-ft.-wide section, and removed it. This approach gave us access to the attic without having to go through the house and also provided light and ventilation to the attic where we would be working.

After the new rafters were securely in place, the sheathing and the #30 felt were installed. The next day, we repeated the performance for the rest of the dormer roof. We now had a reasonably weathertight lid on the work area and could begin the demolition process. To keep out rain, we protected the work site with tarps, which were layered like roofing shingles and held fast with furring strips (see the bottom drawing on p. 88).

Keeping a Project Neat

To protect the house from dust and debris, the occupied space was sealed with drywall and plastic. Also, the debris was removed (and new materials brought in) through the new dormer windows.

Construction Sequence Minimizes Disruption

1. Evicting the owners from their bedroom was the most intrusive part of the renovation, so this room was finished first. The outside walls here were completed and the trim in place before demolition started elsewhere.

2. The new closet wall was framed and drywalled (including the doorway) on the occupied side, creating a solid barrier to construction fallout. Plastic sheeting would not have sealed the walls nearly as well.

3. This closet was finished quickly. The bathroom would take the most time and require the most schedule-juggling, which is why it was planned to be done last.

4. Toward the end of the project, the window opening in the bathroom's outside wall was the only access to the work area. The job was finished working toward this opening, with the window installed and the bathroom paint and trim finished just before the new work was opened to the rest of the house.

Cutting tools were set up on the scaffolding outside to avoid noise and mess inside.

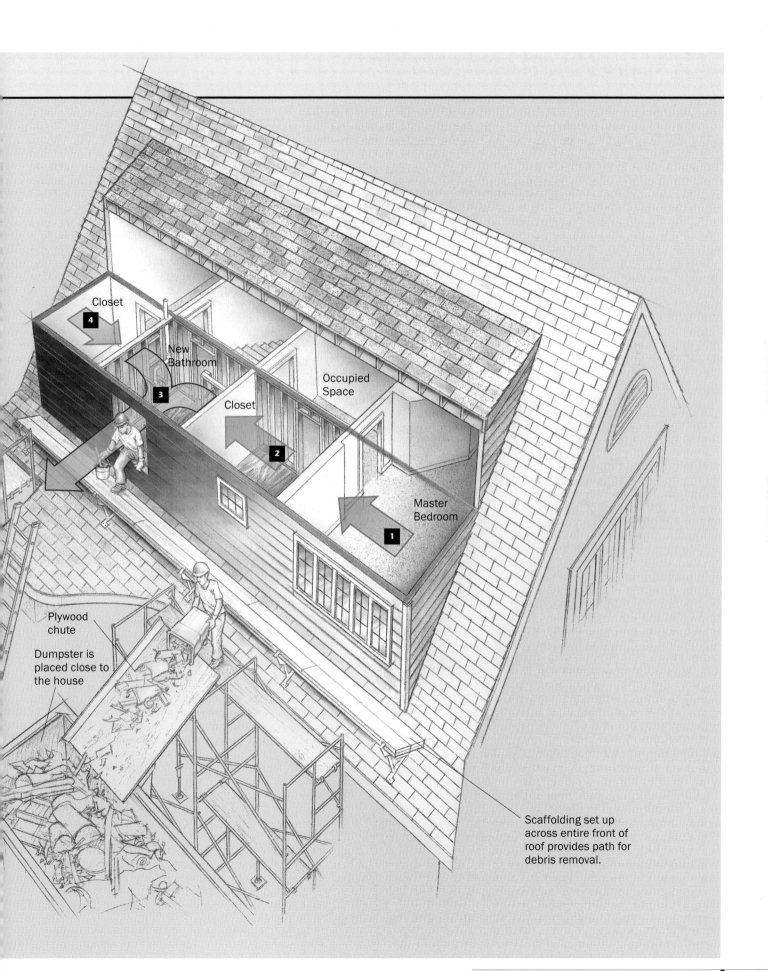

Closet

4

New
Bathroom

3

Closet

2

Occupied
Space

Master
Bedroom

1

Plywood
chute

Dumpster is
placed close to
the house

Scaffolding set up
across entire front of
roof provides path for
debris removal.

Staying ahead of the Weather

To keep out rain, build the new roof before removing the old one. Supporting the new dormer roof with one long header and four posts meant cutting only four small holes in the old roof.

1. Setting and temporarily flashing the posts was the first day's work. **2.** Half of the dormer's new roof was framed and then sheathed the next day, and all the openings were flashed and tarped against the wind and rain. **3.** The dormer's roof was completed and sheathed, and #30 felt paper kept out the rain.

DUCT TAPE AND PLASTIC FOR TEMPORARY FLASHING

Flashing is slipped under the shingles, bent and canted at an angle to divert water around the post. Heavy plastic sheeting (6-mil poly) is slid under the flashing, wrapped around the post, and tacked down with roofing nails. Duct tape seals the poly to the posts.

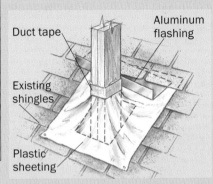

Duct tape

Aluminum flashing

Existing shingles

Plastic sheeting

HAVE TARPS IN PLACE BEFORE IT RAINS

If the tarp is in the way of the work, then secure the top edge to the roof, roll up the tarp around a 1x3 furring strip, and tie the rolled-up tarp in place. When the rain starts, cut the ties.

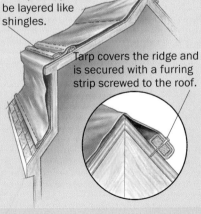

Multiple tarps can be layered like shingles.

Tarp covers the ridge and is secured with a furring strip screwed to the roof.

Keep the Work Site Separate from the House

The interior of the dormer was divided into four areas: master bedroom, closet, the daughter's bathroom, and her closet. It was important to complete the master bedroom quickly, so we built the front wall, sheathed it, installed the window, and had it drywalled and trimmed before some of the other sections were even started. By finishing this room first, the owners got their bedroom back quickly.

We were careful to isolate the workspace from the rest of the house. Two small existing closets were turned into walk-in closets (see the drawing on p. 87). Before we commenced with serious demolition of either closet, we reframed the door openings, drywalled over them, and taped the seams, planning to cut out the doorways after the dirty work was done. Having the living side of the wall closed acts as a barrier against construction while allowing the electricians and plumbers to do their work on the other side.

Even the small amount of drywall we did on the living side of the wall meant dust. We took the usual precautions. A plastic runner protected the upstairs carpet. To contain airborne dust, I tacked tarps across doorways, placed a box fan in a window as an exhaust, and vacuumed at the end of each workday. On the other side, we completed almost all the work before we cut through to install the new doors and trim.

Because the bathroom door wasn't being moved, we taped it shut and put a sheet of foam insulation over it for protection. The bathroom then was gutted and rebuilt.

No Muss, No Fuss

Much of this job revolved around not making a mess. We planned the placement of the Dumpster so that we could build a ply-wood chute to it for debris removal, but kept it out of the way of deliveries. We were working directly over the driveway, and we made sure we swept up carefully every night.

The bathroom's outside wall was left unframed as an access to the second floor, and consequently, little traffic went through the house. The dormer was finished toward that opening, and the day before the tile was installed in the bathroom, we finished framing the rough opening and hanging the drywall, then installed the window and trim. The tilesetter's wet saw was set up on the scaffold outside, and his helper spent the day getting a tan and passing cut tiles through the window.

Every remodeling project is an intrusive ordeal for the homeowner. The single most important aspect of our remodeling strategy was that my clients appreciated the efforts we took to protect their home and privacy.

Nicholas Pitz is the principal of Catamount Design & Construction, Inc. He, his wife, and his cat reside in suburban Philadelphia.

(see the drawing on p. 87)

TIPS

Here are some helpful hints for any remodel.
- *Minimize traffic through the house.*
- *Scrupulously seal off the living space from the work zone.*
- *Use a window fan in the work zone to exhaust dust and fumes.*
- *Take extra precautions to protect the stuff that stays. If the floor stays, for instance, cover it with plywood (and tape the seams) rather than rely on a tarp.*
- *Clean up the work zone at the end of every day.*
- *To keep peace in the house, consider hiring a cleaning service halfway through the project.*

With the large dormer complete, all that remains is adding siding to the gable end wall, which received a new window, to match the dormer windows.

Framing a Dramatic Dormer

■ BY JOHN SPIER

A lot of houses built on New England's coast use dormers to tuck light, airy living spaces under their roofs. Steep-roofed A-dormers are an attractive approach to this style. A house my crew and I recently built on Block Island, Rhode Island, incorporated three A-dormers on the front of a conventional colonial-style roof.

What Is an A-Dormer?

An A-dormer differs from most other dormers in that its gable wall is built in the same plane as the exterior wall of the house below it (see the photo on the facing page). It's just about the only dormer that looks good when not recessed into the roof. With steeply pitched roofs (these were 24-in-12), A-dormers provide minimal floor area, but they can accommodate tall windows and interesting cathedral-ceiling details (see the photo at left). The exterior trim is usually simple with a uniform soffit width and long rakes connected by minimal lengths of horizontal fascia.

Because the entire A-dormer roof extends down to form a valley with the main roof of the house, none of the usual dormer details, such as cheeks, corner boards or siding-to-roof flashings, is an issue. Absence of these details makes A-dormers easy to finish and weatherproof.

First Building Steps Are the Same as for a Doghouse Dormer

As is the case with most other dormers, the main roof of the house is framed first: The gables are raised, the ridge beam is set, and the common rafters are installed with double or triple rafters framing the openings left for the dormers. The dormers for this project were supported by 5-ft.-high kneewalls that extend inward from the outside wall; in turn, beams in the ceiling below carry the kneewalls. We framed these kneewalls first and set all the common rafters for the main roof outside of them.

Because A-dormer gable walls are flush with the wall below, the adjacent common rafters must have their tails cut off flush with the exterior-wall plate (see "Not Your Basic Dormer" on p. 92). A triangulation using the dormer pitch and the kneewall height showed which common rafters needed to be cut flush and which needed full tails to carry the fascia between the dormers.

Next, we framed the gable faces of the dormers. Lines snapped on the subfloor represent the top and bottom plates and the king studs, and the walls are framed to the snapped lines and lifted into place in a miniature version of raising the main gable.

The 24-in-12 pitch of these dormers required bevel cuts of 64° for the tops of the studs, which is beyond the reach of any saw I own. Because each of these narrow dormers had about only eight of these cuts, I mass-produced the cuts by stacking the studs on edge and gang-cutting them. An 8¼-in. circular saw helped, but a regular saw with a reciprocating saw to finish the cuts also would work.

Next, we took the height of the dormer ridges from the plans and installed headers between the doubled common rafters on each side of the dormers. We nailed in hangers to support

the ends of the headers. After the gables were made plumb, the ridge boards were cut, laid out and nailed in place. Next, we made and tested a pattern rafter (without a tail), and we set a pair of rafters on each dormer gable. We cut and installed short studs under these gable rafters outside the kneewalls as nailers for the sheathing and as a way to tie the common-rafter ends to the gable rafter of the dormer. After the rafters were in place, we extended the sheathing from the main house onto the dormer gable.

Two Types of Valleys in Each Dormer

Roof valleys can be framed two ways, and each dormer valley in this project is a hybrid of both. The first method uses a valley rafter, which carries jack rafters on each side. This approach was used for the valleys above and inside the plane of the kneewalls where the interior was to be finished as cathedral space.

Not Your Basic Dormer

Although dramatically different when finished, an A-dormer's initial framing looks much the same as a doghouse dormer **1** with kneewalls and a small gable wall. But then the gable rafters extend down to meet the main roof of the house **2**, and the unique A-dormer character begins to emerge. The wall sheathing of the main house continues onto the face of the A-dormer **3**.

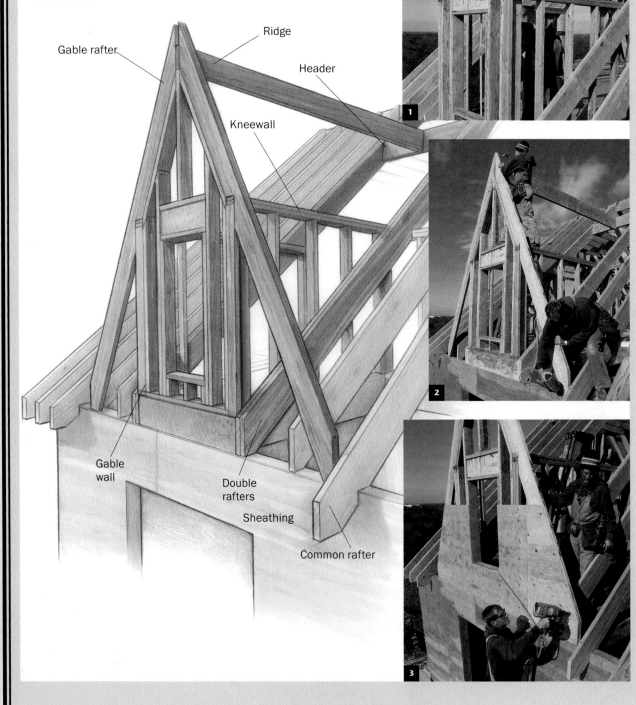

Gable rafter

Ridge

Header

Kneewall

Gable wall

Double rafters

Sheathing

Common rafter

Gathering Valley-Rafter Information

To find the bevel angle for the valley rafters, the author plumbs down from the ridge **1**. Lines drawn from the corner of the kneewall represent the valley, and a rafter square finds the angle **2**. That angle then is transferred to the top of the kneewall **3**, and the length of the seat cut is measured for the valley-rafter layout **4**. A string from the kneewall to the ridge is used to find the plumb cut **5**.

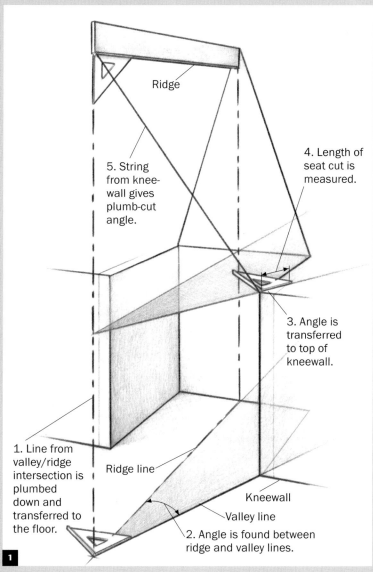

Ridge

5. String from knee-wall gives plumb-cut angle.

4. Length of seat cut is measured.

3. Angle is transferred to top of kneewall.

1. Line from valley/ridge intersection is plumbed down and transferred to the floor.

Ridge line

Kneewall

Valley line

2. Angle is found between ridge and valley lines.

Cutting the Valley Rafter

After the ends of the valley rafters are laid out, the first step to cutting the steep bevel angle is making the heel cut or plumb cut at the complementary angle, in this case 18° **1**. With the board on edge, a saw set at 90° rides on the first cut to create the bevel angle of 72° **2**. A reciprocating saw finishes the cut **3**.

1. Complementary angle of heel cut (18°) is made.

Seat cut is made at 90°.

2. Saw set at 90° rides on first heel cut for a 72° angle.

Flip the board on edge.

3. Finish the heel cut with a reciprocating saw.

The second approach, sometimes known as a California valley, was used for the valleys below and outside the plane of the kneewalls where the interior space was unimportant. A California valley is made by framing (and in this case, sheathing) the roof on one side of the valley, then building the roof for the other side of the valley on top of it.

Valley Rafters for Tops of Dormers

Any regular valley between roofs of equal pitch is fairly straightforward; the compound cuts are all beveled at 45°, and standard tool settings and references are used to calculate the valley-rafter angles. An irregular valley, which joins roofs of different pitch, introduces several complications. These A-dormers with their 24-in-12 pitch intersected a roof with a 10-in-12 pitch, creating an extremely irregular valley.

The first step is finding the angles and location for the valley rafter (see "Gathering Valley-Rafter Information" on p. 93), which is possible through advanced mathematics. But every carpenter I know does it by snapping lines and taking measurements. I start by transferring the location of the valley/ridge

Just as if it grew here. When the valley is in place, the 72° heel cut lies flat against the doubled common rafters. The rafter bottoms are kept flush, and the dormer roof sheathing hides the height difference in the rafters.

Building a California Valley

With a standard valley above and the main roof sheathed **1**, the crew snaps a chalkline for the edge of the California valley **2** and measures the length. A 2x plate with a beveled edge forms the base of the valley **3**, and the rafters cut to length land on the plate **4**. Sheathing is installed from the top down **5**.

Challenging but dramatic. The steep pitches and sharp angles of A-dormers are an extra challenge for the roofing plastering and painting subcontractors, but the dramatic results are worth the extra effort.

intersection to the floor with a plumb bob or a tall level. With a straightedge, I draw a line from that point to the corner of the kneewall, which supports the lower end of the valley rafter. Using a rafter square, I then measure the bevel angle directly from the floor.

Next, I stretch a string from the corner of the kneewall to the ridge and record the angle of the plumb cut with a rafter square. The length of the valley rafter then is measured between those points (along the bottom edge of the rafter). I transfer the bevel angle of the valley to the top of the kneewall, and then I can measure the length of the seat cut to complete the valley-rafter layout.

Another complication when dealing with irregular valleys is the different depths of the rafters for the different roof pitches. In some cases, the width of the stock can be adjusted. In this more extreme case, I framed everything to the planes of the interior ceiling, then allowed the plane of the deeper dormer roof to land beyond the valley rafter on the roof sheathing of the main house.

One of the greatest challenges to framing these dormers is cutting the extreme compound bevels on the ends of the valley and jack rafters (see "Cutting the Valley Rafter" on p. 94), in this case at 72°. After laying out the cutlines, I cut the complementary angle (90° minus 72°, or 18°) along the heel of the rafter. Then I stood the rafter on edge and made a 90° cut, letting the saw ride along the bevel cut I just made, and I finished the cut with a reciprocating saw.

California Valleys Simplify Framing the Lower Dormer Roof

After the valley rafters and jack rafters are installed, we lay out and finish the rest of the roof. First, we make a set of fly rafters with blocks to establish the overhang. These rafters are installed with the lower ends left long, and they then are cut in place to the level of the common-rafter tails

from the main roof. Next, we install short sections of subfascia to join the fly rafters of adjacent dormers.

As I mentioned, we sheathe the common roof with plywood before installing the plate for the California valley (see "Building a California Valley" on p. 95). For strength and simplicity, the sheathing runs to the sides of the dormer kneewalls.

We locate the plate for each California valley by snapping a chalkline from the valley rafter to the outside edge of the gable rafter. After measuring the length, we cut a 2x10 plate (wide enough to catch the tails of the rafters) that forms the base for the California valley. We bevel one edge of the plate to the angle of the dormer roof and cut the angles for the ends, which we figure using a rafter square along our snapped line. We nail the plate to the snapped line. The rafters for the California valley are identical from the bird's mouth at the kneewall up, so it's just a matter of cutting the tails to length.

Top-Down Sheathing Removes the Guesswork

The last step is sheathing the dormer roof. As with most dormers, and especially with roofs this steep, I find it easiest to work from the top to the bottom. Not only do I usually have safer, more comfortable footing with this method, but measuring the remaining angled pieces is also easier.

After snapping the course lines, I start with a full sheet of plywood at the upper and outermost corner of the dormer roof. The trapezoidal shape of succeeding pieces of plywood then can be given to the person doing the cutting simply by measuring the short and long points or by using overall lengths and the common difference.

*John **Spier** and his wife, Kerri, own Spier Construction, a custom-home building company on Block Island, Rhode Island. His book,* Building with Engineered Lumber, *is forthcoming from The Taunton Press.*

Dramatic Skylight

■ BY MIKE GUERTIN

Skylight wells, or shafts, usually are boring. The straight-sided garden-variety shaft isn't much larger than the roof window itself. Plus, the small opening limits the light it lets in. Whenever it's feasible, I splay all four of the shaft walls and flood the whole room with natural light (see the photo below).

Just because you've got a truss roof doesn't mean you can't have a huge skylight opening in the ceiling. Painted white, the truss members become a striking architectural feature as they pierce the light-filled opening.

Dealing with the Truss Dilemma

Building a skylight shaft is tricky in a truss roof. Building codes prohibit modifying the webs and chords of engineered trusses in the field without an engineer's approved design. I prefer to leave the trusses and any bracing intact, with the webs and chords exposed as they run through the opening. When painted, the trusses add character to the opening. And I don't have to modify the roof structure. For this project, I installed a pair of skylights spaced one truss bay apart (24 in.). The sides angle to the adjacent trusses on each side.

Lay Out the Opening before the Skylights Go In

I lay out the opening in the ceiling before locating and cutting the skylights in the roof. Keeping in mind which truss bays the skylights are going into, I poke holes through the ceiling to mark the inside edges of the trusses next to those bays. I snap chalklines through the holes to locate the sides of the skylight's splayed opening.

Next, I decide how much of the existing flat ceiling I want to leave beyond the top and bottom walls of the splay. With the corners established, I snap chalklines to complete the layout for the opening (see the photos at right and on the facing page) and drill a hole through the ceiling at each corner to transfer its position to the attic. Now I can determine the location of the skylights.

To contain the mess, I do as much dirty attic work as possible before opening the ceiling. I remove the insulation from the area, install the skylights in the roof, and give the trusses that will be exposed a quick coat of primer. The final bit of prep is covering the floor and walls of the room with sheet plastic to contain the dust and debris from the skylight well. To collect the dust from cutting the drywall, I hold a vacuum

hose next to the reciprocating-saw blade. I follow the snapped line as closely as possible to make the drywall taping easier. When cutting the lines perpendicular to the trusses, I stop cutting before I hit the edges of the truss chord.

Each Side Rafter Is a Different Angle

After the drywall is removed from the opening, it's time to start the framing that supports the insulation and drywall. The skylight well sort of resembles a large ceiling coffer, except the angle of the roof turns the sides into irregular four-sided shapes. For the sake of simplicity, I'll refer to the framing supports as rafters. The rafters for the splayed sides are each a different length, and each rafter sits at a different angle. So the angle of the end cuts changes as well.

I use two strings to locate the first rafter position and to establish its cut angle. The first string is plumb from a lower outside corner of the framed skylight opening to the bottom truss chord. From that point, I square over to the truss chord at the side of

the opening where the rafter lands. I then run a second string from that point back to the corner of the skylight. To secure the string at the lower point, I slip a dull putty knife between the drywall and the bottom truss chord. I now can measure the length along the diagonal string as well as the angle of the cut (see "Skylight-Well Framing" on pp. 100–101).

The rafter's bottom cut is a straight plumb cut, but the top cut is a bird's mouth with the seat cut angled to the roof pitch. I cut the rafter for one side and test it. If I'm satisfied, I cut its mirror image for the other side and screw them both into place. Having established a starting point, I reposition the strings for the next set of rafters either 12 in. or 16 in. away, depending on the size of the skylight. The process is repeated for each pair of rafters until I reach the upper corner of the skylight (see the center photo on p. 100).

Corner Rafters Are Really Hips

The corner rafters are like hips that reach from the corner of the skylight to the corner of the ceiling opening. Each hip cut is compound with both a bevel angle and a miter angle. I stretch a string from the skylight corner to the corner of the opening to determine the miter angle. To figure out the bevel angle, I plumb down from the skylight corner to the bottom truss chord. A string that I run from that point to the corner of the opening gives me the bevel angle. Before fastening the hip rafter in place, I center it on the string.

Just as with a hip roof, I measure, cut, and install any jack rafters, maintaining the same rafter spacing as I work (see the bottom photo on p. 100). The end rafters along the bottom and top of the skylights are equal in length and position; I attach one to the side of each truss. The last step I take before hanging the drywall is installing batt insulation between the rafters.

Templates for Twisted Walls

Because the sides of the skylight well are odd four-sided shapes and have a curving plane, there's really no way to measure,

Open up the ceiling. Snapped lines mark the shaft opening on the ceiling (photo at left). As the opening is cut, a vacuum hose held near the sawblade keeps dust to a minimum (photo at center), and the old ceiling should come down in large pieces that leave less mess (photo at right).

Skylight-Well Framing

To frame the skylight shaft, 2x4 rafters connect the top and bottom chords of adjacent trusses (see the drawing on the facing page). Because the roof plane slopes up and away from the ceiling plane, each side rafter has a different length and angle. As the rafter lengths increase, the cut angles increase as well, creating a twist in the sidewall. Hip rafters form the corners, and jacks complete the sidewall framing.

Strings find the length and angle. A plumb line squared over to the side of the opening locates the first rafter (photo left, point A on drawing). A string back to the skylight provides the length and the proper angles, and the rafter then is screwed into place (photo right).

No two rafters the same. Measuring off the first rafter locates each additional rafter whose length increases as the angle to the skylight becomes steeper (point B on drawing).

Hip corners. With all the rafters in place, the twist in the sidewalls is unmistakable. Hip rafters form the corners of the shaft, and jacks complete the framing (point C on drawing).

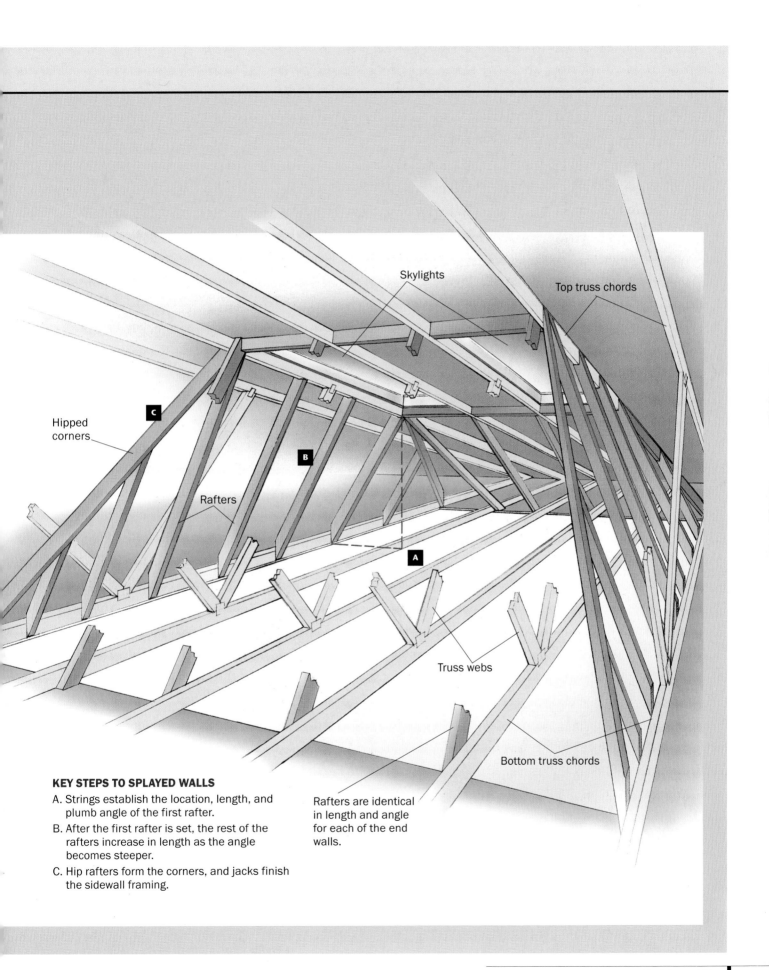

Skylights

Top truss chords

Hipped
corners

C

B

Rafters

A

Truss webs

Bottom truss chords

KEY STEPS TO SPLAYED WALLS

A. Strings establish the location, length, and
   plumb angle of the first rafter.

B. After the first rafter is set, the rest of the
   rafters increase in length as the angle
   becomes steeper.

C. Hip rafters form the corners, and jacks finish
   the sidewall framing.

Rafters are identical
in length and angle
for each of the end
walls.

Fit Drywall in the Skylight Well

Because the sidewalls of the skylight are not flat planes, the most accurate way to copy the exact shape of the wall is with a plywood-strip template. The template transfers the shape to a sheet of drywall, and the drywall is dampened slightly so that it bends into the twisted plane of the sidewalls.

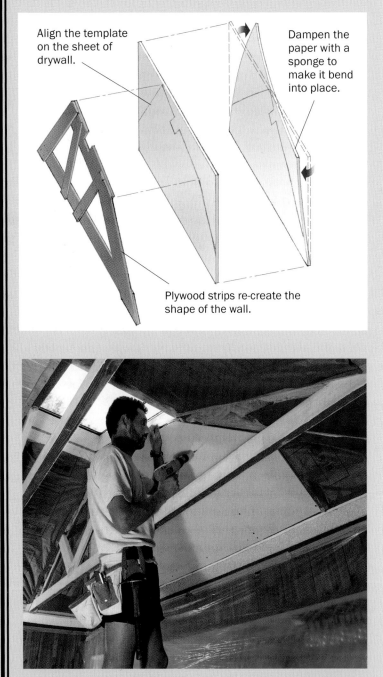

Align the template on the sheet of drywall.

Dampen the paper with a sponge to make it bend into place.

Plywood strips re-create the shape of the wall.

Screws and a little water bend the drywall. To bend the dampened drywall to the shape of the sidewall, fasten the bottom edge to the framing first, then drive many closely spaced screws at each rafter location to shape the drywall a little bit at a time.

mark, and cut the drywall without a template. So I make the template out of strips of ¼-in. plywood, which bends easily.

I first cut the strips into rough lengths for the edges of each side and screw them to the framing temporarily. Next, I screw together the corners of the template strips (see "Fit Drywall in the Skylight Well" at left). I rough-cut openings around any obstructions, such as the truss webs, and locate the precise dimensions of the cutout onto the template with a scribe block. Later, I use the scribe block to transfer the cutout location to the drywall. To reinforce the template, I attach diagonal braces in a truss-like fashion. Once the template is done, I mark the rafter positions along the perimeter, back out the temporary screws, and then carefully remove the template from the wall.

Water Makes Drywall Flexible

If the template is longer than 8 ft. (as this one is), I place it on a sheet of drywall with a rafter position lined up on the end of the sheet. I trace the template on the drywall and mark the cutouts precisely with the scribe block.

The sides of the skylight well curve, so the drywall needs to bend to the plane. The easiest way to bend drywall is by doubling up ¼-in. sheets, but ¼-in. drywall isn't always available. If ½-in. stock is all I can get, I cut the pieces and wet the paper on both sides of each piece with a damp sponge. I let the sheets sit for about 15 minutes and repeat the sponging. Damp drywall is fragile and heavy, so an extra set of hands is helpful for positioning the sheets.

Lots of Screws Bend the Drywall

To attach the first sheets, I drive screws along the bottom edge at each rafter. Then I work up each rafter, spacing the screws about

Patterns for Odd Shapes

A plywood-strip pattern or template can re-create any oddly shaped area, whether it's a crooked wall inside a closet or the wall of a splayed skylight well. The strips follow the edges of an area and can be scribed to fit edges that aren't straight. For obstructions such as a truss web, the loose shape is cut into the strip, and a scribe block marks the exact location. For this template, rafter locations marked on the template help to orient it properly on the sheet of drywall.

Plywood bends to the wall. First, ¼-in. plywood strips are screwed into the framing. Next, the corners are joined (photo left). Diagonal strips reinforce the template, and then the complete template can be removed from the framing (photo right).

Scribe block. To copy an exact shape onto a larger template, a scribe block is fast and accurate. Butt the block against the shape (in this case, a truss web), and trace the edge of the block (photo left). Then transfer the shape to the finished material (photo right). A cut corner on the block aligns it exactly on the template.

Where the drywall has to fit around the truss webs, smaller pieces are easier to fit. The upper end wall is done in two pieces (photo below), and the lower end wall can be made in one piece. Triangular pieces complete the board installation at the corners.

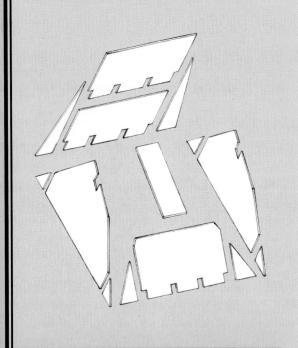

8 in. apart where the sheet is nearly flat and 4 in. apart or less at the most severe curves.

When ½-in. drywall is bent, the paper face almost always splits slightly during the process, but any damage can be fixed later with compound. Smaller drywall pieces fill in the ends of the sidewalls. With the drywall installed, the curve of the wall becomes pronounced.

Drywalling the top and bottom walls of the skylight well is simpler than drywalling the sidewalls. Because the planes are flat and the edges regular, I can measure and cut notches around the chords and webs instead of making templates. Where truss webs intersect the upper end wall, I install two sheets horizontally for one long seam rather than cut separate pieces to fit between the trusses (see "Patterns for Odd Shapes" on p. 103). The lower end wall can go in as one piece.

Mud, Body Filler, and a Little Paint

The flat ceiling panel between the skylights is filled with rigid-foam insulation and then drywalled over. I also drywall the inside edges of the skylight curbs. To finish around the truss framing where it intersects the top and bottom of the well, I apply tape and joint compound. I tape and compound over any splits in the paper on the curved sidewalls as well. It takes several coats to fair over the joints and any splits. With all the picky little spots and the curved surface, it might be wise to leave the mudding to a pro.

To give the trusses a more finished look, I cover the truss plates with auto-body filler, feathering the edges back to the 2x4 truss webs and chords. After they're painted, the trusses have a nice, smooth look.

Mike Guertin*, a contributing editor to* Fine Homebuilding, *lives and works in East Greenwich, Rhode Island. He is the co-author of* Precision Framing *(The Taunton Press, 2001) and the author of* Roofing with Asphalt Shingles *(The Taunton Press, 2002).*

Skylight Kitchen

■ **GEORGE BURMAN AND PATRICIA LOONEY-BURMAN**

Plenty of light. A flared skylight well lets daylight reach just about every corner of this kitchen. Photo taken at A on the floor plan (p. 106).

ark wood cabinets, brown and orange vinyl floor, avocado appliances, yellow tile counters, and a four-tube fluorescent light box. Those were the highlights of the kitchen in the house that we bought 12 years ago in California's San Joaquin Valley. We loved the house, set on a 1-acre lot in a semirural area with mature landscaping and a feeling of spaciousness. But that kitchen had to go.

We made a pass at it a few years later with a new paint job, some new appliances and a small island with a granite top. But this interim kitchen was just that. We still needed space for two cooks, more daylight, cabinet doors that were not dirt collectors, and work surfaces that were beautiful and functional.

Assembling a Team the New-Fashioned Way

Having decided to go for a full-blown remodel, we contacted an interior designer we found through www.improvenet.com, a service we learned about in *Fine Home-building*. We had only one response to our ImproveNet™ query, Marlene Chargin. But

A Kitchen with a Sky-Lighted Island

Annexing a little more than 100 sq. ft. of the backyard patio expanded this kitchen into a room large enough to include a breakfast table and a door to the patio. And placing a 2-ft. by 6-ft. skylight over this kitchen's island ensured plenty of day-lighted workspace for two cooks.

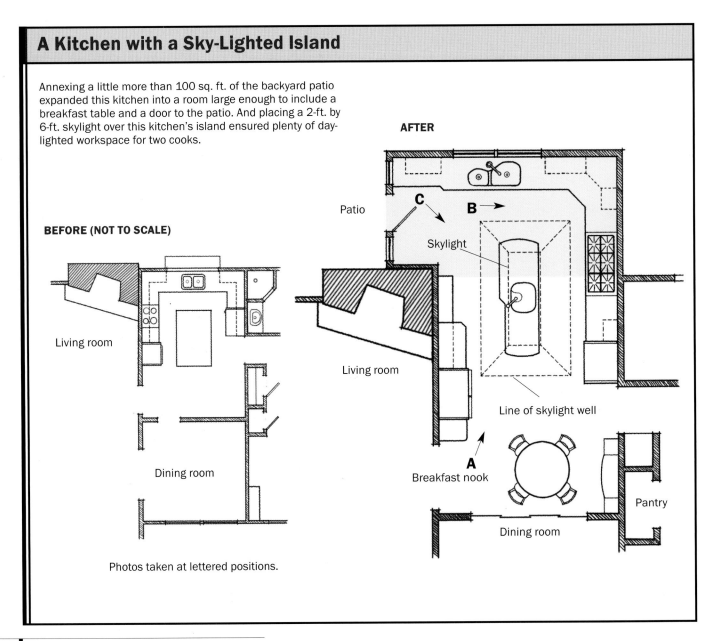

BEFORE (NOT TO SCALE)

Living room

Dining room

Photos taken at lettered positions.

AFTER

Patio

Living room

Skylight

Line of skylight well

Breakfast nook

Dining room

Pantry

you need only one if it's the right one. Mar-lene referred us to several of her clients and their kitchen projects. We were impressed with the work and with the other members of her team: kitchen designer DeAnn Martin and builder Mark Fletcher. Reassured by the references, we signed up this trio to direct our remodel.

Marlene and DeAnn listened to what we wanted, steered us clear of some pitfalls and made us feel that the proposed kitchen would really be "our" kitchen. The builder also communicated with us and with the designers well, keeping us informed of daily progress and anticipating design and construction problems before they required reworking.

Clean and simple. Thermofoil coatings are plastic laminates that stretch around corners and edges to minimize dirt-catching seams. This piece, shown with its protective film peeled back, will be a drawer front.

Easy Decisions, Hard Decisions

We easily settled on a layout that involved bumping out the existing kitchen into a covered patio area (see the floor plans on p. 106). This plan added about 112 sq. ft. to give us a total of 320 sq. ft. for the new kitchen—not huge, but comfortable for two cooks.

A 2-ft. by 6-ft. skylight was also an easy decision. It is centered over the island in a splayed opening that lets daylight into the entire room (see the photo on p. 105). Made by Distinctive Skylights, the skylight includes a double layer of translucent fiberglass panels separated by a grid (see the photo at right). The grid pattern imparts an unexpected design element, the fiberglass panels provide some thermal insulation, and the light is softened by their translucence.

Granite countertops (Blue Pearl from Norway) and tile for the floor were natural choices. Reusing the refrigerator, dishwasher, and dual ovens was also an easy decision. We had a much harder time choosing the cabinetry style and finish.

Our neighborhood is surrounded on three sides by orchards, so dust is a chronic problem in our house. Raised-panel doors

Skylight softener. Two layers of fiberglass separated by a grid cut down on thermal gain through the skylight and soften the light falling on the kitchen surfaces.

A splash of brilliant blue. Etched-glass tiles spread out from the center of the cooktop, amplifying the highlights that flicker out of the granite counters. Photos taken at B on the floor plan (p. 106).

with crevices and corners are difficult to keep clean here. Besides, we prefer a sleek, contemporary look. Marlene suggested rigid Thermofoil (RTF), a heat-formed laminate that wraps around the corners of cabinet parts. We went to the local home-improvement showrooms to look them over and saw samples that looked cheap and tacky. Then Marlene located some RTF samples by Premoule of Ontario, Canada. One of their finishes, Silver Ash, was light and silvery, with a subtle wood-grain pattern we loved.

We went about securing a bid for our job from an out-of-town company that specializes in Premoule cabinets. Their price came in at 75% more than our local cabinet shop

wanted for wood cabinets. So we went to plan B. Our local shop farmed out the fabrication of all doors, drawer fronts, and exposed cabinet panels directly to Premoule. This persistence paid off when the cabinets came in at the original bid price for wood.

The result is a light, clean look. The Thermofoil material is seamless, even at the corners, making for easy cleanup (see the top photo on p. 107).

Glass in Unexpected Places

The challenge in the cooking area was the treatment of the backsplash. Marlene suggested glass of some sort. Plate glass, either

Now there's room for a breakfast table. Bumping the kitchen towards the backyard allowed the island to move away from the dining room. In the foreground, a custom-fit chopping block goes over the sink opening. Photo taken at C on the floor plan (p. 106).

Sources

Island lighting
Tech Lighting
MonoRail
(847) 410-4400
www.techlighting.com

Skylight
Distinctive
Skylights
(800) 430-0076

Thermofoil cabinet
components
Premoule
(866) 652-1422

Wall tile
Ann Sacks
Azure by Lake Garda
(800) 278-8453
www.annsacks.com

Is a Warming Oven a Useful Appliance or Just Another Electronic Gadget?

We hadn't even thought of a warming oven until we were at the appliance dealer's one day and asked about it. The dealer described it, adding that clients who had one said they couldn't live without it. After checking with the kitchen designer and finding that the island could accommodate it, we ordered one. It has been great.

At first we thought we'd probably use it just at Thanksgiving, when we're try-

Warming oven fits in a drawer.

ing to get all the hot things served at the same time. We're pleased to report that it serves that function just fine. But we've found plenty of other duties for the drawer. We use it for warming plates (frequently) and are also using it as a place for yeast dough to rise, an unexpected bonus. It also makes a good defroster—not as fast as a microwave, but it won't start to cook the food, either. Made by Dacor® (800-793-0093), the warming oven has a range of 90°F to 200°F. It doesn't use much electricity (4 amps at 120v) and cost us $762. The appliance dealer was right.

glossy or sandblasted, was ruled out because it is too hard to keep clean. We finally settled on etched (more of a satin finish) glass tiles. They passed the grease-cleanup test, which was to apply no-stick cooking spray to the tile, then clean with a glass cleaner to see if there was any residue or streaking. With sandblasted glass, some of the oil remained; glossy glass had streaks; etched-glass tile cleaned just fine.

A stepped pattern with small silver glass-tile accents complemented the shape and finish of the exhaust hood (see the photos on p. 108). At the centerline of the stove, the bookmatched tile patterns come together at a strip of granite that matches the counters.

The island also presented a problem that wasn't resolved until late in the project. We had originally planned to install two appliances in the island: a warming oven (see the sidebar above); and a Kohler® combination prep sink/pasta cooker. The shipment of this component was delayed

several times until after the island granite top was temporarily installed. At that point, we realized that the Kohler sink/cooker would eat up more of the island surface than we wanted. We were able to cancel the Kohler order and substituted a smaller prep sink. A pot-filler faucet over the range top replaced the pasta cooker.

We got a cutting board at the local kitchen-supply shop and shaped it to fit the island's sink opening (see the photo on p. 109). That extended the island's counter space so that we can work without getting in each other's way.

Price estimates noted are from 2001.

George Burman and Patricia Looney-Burman cook *together at their home in Madera, California.*

Shedding Light on Skylights

■ BY ROE A. OSBORN

Skylights for different purposes. Skylights on the outside of this house punctuate the various roof planes. Inside, they serve a variety of functions, from kitchen light and ventilation, to ventilation and added views for a second-floor bedroom, to concentrated light and solar gain in the sunroom.

After 10 years my wife and I had almost gotten used to the dark, windowless crypt that served as our bathroom. There wasn't enough money in our original budget for a bathroom skylight, so we added it to our wish list, somewhere in between a new microwave and a garage. But after finding the right skylight at a sale, we moved the project up the list.

A skylight expands a bathroom. The splayed chase for this skylight not only lets in more light but also makes the bathroom feel roomier. Even though the bathroom is equipped with a powerful exhaust fan, an open skylight can provide a quick escape route for the warm, moist air from the shower and the spa.

Using an old table lamp without a shade for light, I cut the hole in the ceiling and did the necessary framing alterations in the dim and dusty roof cavity. Then, on a steamy July morning, I stripped the roof shingles back and blasted through the sheathing, plunge-cutting with my sidewinder.

When I lifted out the rectangle of sheathing, a blast of air from inside the house blew sawdust everywhere. The change in the bathroom was dramatic. Even without the chase closed in, the bathroom went from dreary to cheery as sunlight flooded in. A refreshing, gentle breeze wafted through the room as the natural convection currents inside the house kicked in.

To Vent or Not to Vent

Besides letting in light, our bathroom skylight was a way to get rid of excess moisture. One of the first things skylight buyers should consider is where the skylight will be going and if it will be needed for ventilation (see the photo on p. 111). If the skylight is going into a kitchen or a bathroom as ours was, a venting skylight is a wise choice. Even with exhaust fans and range hoods, a skylight can provide a quick alternate escape route for the excess moisture and warm air that these areas are likely to encounter (see the photo at left).

Nearly every residential-skylight manufacturer offers the choice between skylights that open to provide ventilation and skylights that are fixed, or nonopening. Bart Mosser, vice president of Wasco® (P. O. Box 351, Sanford, ME 04073; 800-388-0293), told me that Wasco's fixed residential skylights typically outsell their operating skylights almost 3 to 1, probably because operating skylights almost always carry a bigger price tag. The venting version of their E-Class 22-in. by 46-in. skylight lists for $477 compared with $277 for the fixed.

Skylights in high-moisture areas such as bathrooms and kitchens are more likely to suffer condensation than skylights in other areas. Because venting skylights open to allow moist air to escape, they are best-suited for these high-moisture areas. If the condensation is heavy enough, it can run down the glass, over the skylight frame and down the skylight chase, damaging everything in its path.

Many manufacturers include condensation gutters on their skylights to catch and to collect condensation as it runs off the glass. If you are putting a skylight in an area likely to see a lot of moisture, make sure the skylight you choose has these gutters.

Still not convinced that you need a venting skylight? The folks at Velux®-America (800-888-3589) have come up with a compromise. Their FSF skylight has a ventilation flap at the top of a fixed skylight (see the photo at right). The flap opens into a channel to the outside, which allows air circulation even in bad weather. The FSF skylight adds only about $30 to the cost of their FS fixed skylight of comparable size and glazing, a reasonable price for convenient ventilation. Velux, however, still recommends fully venting skylights for use in kitchens or bathrooms.

Motor-Driven Skylights Operate at the Press of a Button

If a skylight is being installed in a living space inside the geometry of the roof, such as a finished attic space, and you have easy access for opening the skylight, you may opt for one that closes with a latch and pushes open. Other skylights crank open and shut like awning windows. Push-open skylights open wider than their crank-out cousins and usually pivot for easy cleaning of either side.

If your skylight is going in a less accessible spot—say, in a cathedral ceiling high above the floor—then the crank-out variety is probably a better option. Many crank-out skylights also pivot for cleaning, but the opening mechanism has to be disconnected

Fixed skylight with ventilation. Velux's FSF skylight does not open, but a ventilation flap at the top lets warm air out and cool air in through a screened channel.

from the sash beforehand, not an easy feat when the skylight is out of reach. Instead of a handle, many crank-open skylights are equipped either with a socket or a small, fixed loop that lets you operate the skylight with a telescoping crank handle from the floor below.

Most skylight companies offer an optional motor that opens or closes the skylight at the push of a button either from a switch on the wall or a remote control. A lot of these motors look pretty ugly, like large boxes stuck on the skylight trim as an afterthought.

The slickest-looking mechanical skylight opener belongs to Roto's Sunrise II skylights (Roto Frank of America, Inc., Research Park, P. O. Box 599, Chester, CT 06412; 800-243-0893). Roto houses the entire mechanical works in an aluminum extrusion that is wood-veneered to match the skylight trim. The aluminum extrusion conceals both the crank mechanism and the optional motor, and the extrusion can be removed easily for access and maintenance. Motorizing any of Roto's Sunrise II model skylights adds $95 to the price.

A rain sensor waits for rain. When even a single drop of rain hits this small circuit board, it instantly signals the skylight motor to close the skylight and keep out the rain.

The mechanisms that hold crank-out skylights open fall into two categories. One variety has metal arms that swing, or scissor, out to open the skylight. Both Velux and Andersen® (100 4th Ave. N., Bayport, MN 55003-1096; 800-426-4261) use this type of apparatus, similar to the opening mechanisms for awning and casement windows.

The second and most popular system among skylight companies opens the skylight with a chain that uncoils and stiffens as the crank handle is turned. The chains are made either of metal or of plastic, and manufacturers using this system brag that their

skylights open farther than those using the stiff-arm mechanisms.

Smart Skylights Close when It Starts to Rain

When people give me a hard time about being bald, one of my standard comebacks is that you have to be bald to fully appreciate a ride in a convertible. Another advantage to being follicularly challenged is that I'm usually the first to feel raindrops. Some of the contractors I used to work for depended on me for this ability, especially when they were trying to get that last course of shingles on before an approaching thunderstorm.

One thing prospective skylight owners worry about is not being home to close the skylight in case of rain. You could hire a bald guy to sit on the roof and wait for rain, but skylight manufacturers have come up with a better idea: rain sensors, which are small printed circuit boards that look as if they shouldn't be exposed to bad weather (see the photo at left). The rain-sensor circuit board basically consists of two conductors in a grid pattern. When a drop of rain lands on the grid, it completes a circuit that signals a motor to close the skylight.

Virtually every skylight company that offers a motorized opener also offers a rain-sensor option for their operating skylights. The better openers have battery backups for their motors. You may consider battery backup to be overkill, but if the electricity is knocked out and it starts to rain (as often happens in thunderstorms), you'll appreciate battery backup.

Some skylights with motorized openers can also be opened manually in case of emergency. This feature can be important if the motor fails or if you need to open the skylight before the power comes back on.

The controls for automated-skylight functions vary among manufacturers. Some control just the opening and closing functions, and others can be programmed to control motorized-skylight accessories such as the interior shades that most companies offer. Velux says it has a system that will interface with any homewide computer system. The system can control multiple functions on multiple skylights from a single remote box.

Glazing Options Can Reduce UV-Damage and Heat Transferral

Jefferson Kolle, a former renovation contractor, told me that he once dropped a worm-drive circular saw onto a skylight of tempered glass and was amazed when the saw just bounced off. For safety reasons, skylights are made of tempered glass or laminated glass. Tempered glass is extremely strong, but when it breaks, it shatters into a million glass pebbles.

Laminated glass has a thin plastic sheet attached to the glass. It's generally not as strong as tempered glass, but when it breaks, the plastic keeps the glass in a sheet. For this reason, codes in some areas specify laminated glass for the interior-facing skylight glass in certain applications, such as over a bathtub or spa. Skylights with laminated glass usually have tempered glass on the exterior and are designated "tempered over laminated." This option can be ordered for nearly every skylight on the market. It's best to check with a local building official to find out if any stipulations apply to your installation.

Most skylight brochures include performance tables for the different glazing configurations. Those tables are usually broken down into four categories: light transmission, shading coefficient, UV-blockage, and U-value. Light transmission is the amount of light allowed through the glass. Shading coefficient is the amount of solar-heat gain through the glass compared to a single pane of ⅛-in. clear glass. The lower the shading coefficient, the lower the solar-heat gain.

The percentage of the sun's ultraviolet radiation stopped by the glazing is the UV-blockage number. The sun's UV-rays can

Skylight glazing should fit the application. Clear tempered glass was chosen for the skylights in this sunroom to let in plant-friendly light. Skylights that open with a latch are less convenient but were chosen because they open wider for maximum ventilation in summer.

fade and degrade furniture, carpet, and draperies, and they can even discolor wood floors. The last category, U-value, is a measure of heat transfer through a glazing system. The lower the U-value, the better the insulating performance of the glazing system. With seemingly endless combinations and permutations of glass types and coatings, the insulating performance of glass is a topic worthy of a separate chapter. But I'll try to give a brief description of the options that are available.

Like window glass, skylight glass can be coated, such as with a tint or a low-E coating, to affect heat and light transmissions. However, the effectiveness of the various coatings depends on which of the four surfaces in the insulated-glass sandwich are coated.

Southwall Technologies® (800-365-8794) has put a new spin on glazing performance with its Heat Mirror® products, which have a clear film suspended between two sheets of glass. Southwall's Superglass®, arguably the most efficient glazing option for sky-

lights, has two film layers between the glass sheets.

The right combination of glass types and coatings will depend a lot on your situation (see the photo on p. 115). For instance, if you live in warmer climate and you want to put a skylight on a south-facing roof, a tint or a coating that cuts down on solar gain may be more important than the insulating value of the glazing. You'll also want a skylight with a high UV-blockage number. One note of caution: If you choose a skylight with tinted glass, the color of the tint can affect the color of everything lighted by the skylight. The added cost of special coatings and glass configurations varies among skylight manufacturers. The difference in price between Velux's FS 106 unit with clear tempered glass and the same skylight with laminated glass with a low-E coating is only $30, which seems like a bargain given the prospective energy savings over the life of the skylight.

Glazing Performance Is Not the Same for Windows and Skylights

A lot has happened recently in the study of glass performance. The NFRC: National Fenestration Rating Council® (1300 Spring St., Suite 500, Silver Spring, MD 20910; 301-589-6372) certifies the performance of windows, doors and skylights from the major manufacturers with all of their various glazing options. But the numbers provided in the NFRC Certified Products Directory come from testing all of these skylights in a vertical position and may be suspect. When insulated glass is put in a slanted configuration, as with a skylight on a roof, its internal dynamics change dramatically (see the drawings at left). The U-value of certain glazings can degrade up to 30% when changed from vertical to a slope of just 27°.

Insulated-Glass Performance Changes when Installed on a Roof

When insulated glass is moved from a vertical to a sloped orientation, the internal dynamics change dramatically. Between the panes of glass, convection currents cause heat to be transferred from the warmer inside surface to the cooler outside.

1. The convection loop in skylight glazing is short, allowing heat to escape at a much faster rate.

3. The result is that heat loss through skylight glazing is much greater.

2. In a window the convection loop is longer, which slows down heat transfer.

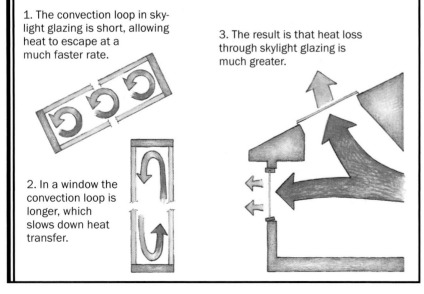

Wide skylights in truss roofs. A truss roof can be engineered to accept a wide skylight. Here, trusses have been doubled, and a small monotruss that fills in the roof below the skylight hangs on a header.

Skylight glazings are available in a mind-boggling variety of coatings and configurations that affect skylight performance. The NFRC is testing each of these variations and combinations in a sloped orientation and compiling new ratings for each skylight manufacturer. Until those new numbers are out, the NFRC's U-value ratings for skylights should be viewed with some skepticism.

Tried-and-true flashing system. Velux's patented system consists of head flashing for the top of the skylight, sill flashing for the lower end and step flashing for the sides. Cladding on the sash seals the top of the flashing from the weather.

Skylights Are Designed to Accommodate Conventional Roof Framing

Skylights have always been made to fit between regularly spaced rafters, either 2 ft. or 16 in. on center. If you're looking to put a skylight between rafters 2 ft. apart, you'll find a variety of skylights with a width of 22 in., usually plus a fraction. If your rafters are 16 in. apart, every manufacturer offers a

A rubber gasket seals the flashing. Many skylight manufacturers use rubber gaskets to cover the top of the flashing, such as the Roto skylight pictured here.

Curb-flashed skylights rely on a mastic seal. When the mastic seal around these curb-flashed skylights failed, roof cement was incorrectly applied on the shingles to stop a leak.

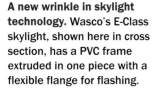

A new wrinkle in skylight technology. Wasco's E-Class skylight, shown here in cross section, has a PVC frame extruded in one piece with a flexible flange for flashing.

skylight that will span two bays, and many offer skylights that will fit into three 16-in. bays or two 2-ft. bays.

Roof trusses can limit skylight options. In new construction trusses can be engineered to accommodate wider skylights (see the top photo on p. 117). But in a remodel you've got to deal with what's there, and altering trusses without an engineer's approval is a no-no. If you are one of the unlucky ones who wants to put a skylight in your 1970s raised ranch but the trusses in your roof are 16 in. on center, take heart. Roto makes the Sweet 16, a fixed skylight that fits into 16-in. on-center truss or rafter bays, as well as a Sweet 16 operating skylight.

Another alternative to cutting rafters is ganging skylights together. Many manufacturers offer flashing kits for side-by-side installations. According to Chuck Silver of Hudson River Design in New Paltz, New York, the best combo kit belongs to Crestline (800-444-1090), which incorporates a raised fin on its skylights to make ganging almost foolproof; they are also good at keeping out the weather.

Flashing Is the Key to a Good Installation

I installed my first skylights back in 1982. The guy I was working for handed me two Velux venting skylights still in the box and told me to prep them for installation. I spent a whole morning removing the aluminum cladding and fastening the mounting brackets at just the right height. Luckily, my boss had installed skylights before, and I watched in awe as he wove in the flashing. My duties were to prevent all of the pieces from blowing away and to keep the screws from jumping off the roof until they could be put back in the cladding. After those skylights were installed, they looked as if they could fend off any kind of weather or an attack from a Klingon squadron.

Aside from the glazing choices, the biggest distinction between different makes of skylights is the flashing. In 1941 in Denmark, V. Kann Rasmussen designed one of the first self-contained skylights. His work evolved into what is now the Velux company. Part of his design was the patented Velux flashing system, which today remains basically unchanged (see the bottom photo on p. 117). One-piece head and sill flashings wrap the top and bottom of the skylight, and step flashing seals the sides. Cladding on the sash, which is removed and replaced during installation, extends down over the top edge of the flashing, serving as counterflashing to create an impenetrable shell.

Other skylight manufacturers such as Roto and Andersen use an EPDM rubber gasket that covers the top of flashing (see the top photo on the facing page). Mike Guertin, a builder in East Greenwich, R. I., says that he prefers Roto skylights because with the gasket, the sash, and all of its cladding do not need to be removed for installation, making the process go a lot quicker.

On the other hand, curb-flashed or perimeter-flashed skylights with a solid flange that runs around the circumference of the skylight always worried me. I'd always associated this type of flashing system with less expensive skylights—that is, until I began my research on this article.

I found many well-made skylights available with welded metal, vinyl, or flexible-PVC flanges (see the center photo on the facing page). The skylights with metal or vinyl have to be installed in mastic to make them waterproof, and the word *mastic* always conjures up images of goo stuck in my beard and sleepless nights as I worried that the mastic seal might not be complete. Nevertheless, many contractors will use nothing else; in fact, these mastic marvels are popular for commercial installations.

The E-Class skylight made by Wasco has a flexible-PVC flange extruded as an integral part of a PVC frame (see the bottom photo on the facing page). The flange has an inverted L about an inch away from the frame; the roof shingles slip under the

inverted L. Outboard of the L are three water-diversion ridges. The design of the flange eliminates the need for mastic or sealant and is self-healing so that it can be nailed without worry. E-Class skylights are carried to the roof in one piece and installed with small metal brackets that lock into a channel in the skylight frame. Installation of a Wasco skylight takes a fraction of the time it usually takes for a step-flashed skylight.

Sun-Tek® (10303 General Drive, Orlando, FL 32824; 800-334-5854) offers a variation on the same theme. Its Elite series uses a welded aluminum flange with similar-looking water-diversion ridges. Predrilled, the flange doubles as a nailing flange and eliminates the need for extra brackets. However, installation of the Elite does require mastic or sealant.

Perimeter-flashed skylights also are installed on top of the roof sheathing, which lets the skylight frame be as large as the framed opening in the roof, while the frames for step-flashed skylights usually fit inside of the roof framing. A larger frame means more glass area and more daylight. Wasco's E-Class 2246 venting skylight has 6.39 sq. ft. of glass compared to 3.99 sq. ft. with Velux's VS 106, or 2.4 sq. ft. more daylight for basically the same-size hole in your ceiling. In all fairness to Velux, as of spring 1997, the company's skylights have featured more streamlined and more installer-friendly flashing that increases glass area for their VS 106 skylight to 4.68 sq. ft.

Special Flashing for Shallow Pitches

The flashing systems I've discussed so far are restricted to installations on roofs with a 4-in-12 pitch or greater. Most skylight companies offer special flashing kits for skylights on shallow pitches, but many of these shallow-slope kits are cumbersome. Because these flashing kits raise the skylight to a higher pitch than the roof, framing and finishing the skylight chase can be a real puzzle. Another alternative for a shallow-pitched roof is a curb-mounted skylight.

Tom O'Brien, a restoration carpenter in Richmond, Va., explained that installing a curb-mounted skylight on a shallow roof (4-pitch or less) involved first building a 2x6 frame or curb on top of the sheathing. Tom usually hires an experienced roofer to fabricate metal flashing around the curb. The flashing is embedded with mastic or sealant, and the roofing material is then run on top of the flashing. The curb-mounted skylight sash with built-in counterflashing is then installed on top of the curb in a bed of mastic or sealant. On a flat roof the roofing membrane is carried up the sides of the curb with the sash mounted on top.

Special flashing kits are also needed if your roof is covered with something other than asphalt or wood shingles, such as metal or tile. If it is, be sure the company that makes the skylight you choose also makes the right flashing kit for your type of roof. This type of installation can be tricky, and I also recommend leaving it to a professional roofer.

Step-Flashed Skylights Are Harder to Make Airtight

Although step-flashing is a great system for keeping water out, its weakest point probably is preventing inside air from escaping. No doubt it must seem strange to consider such a factor when venting skylights are designed specifically to allow air to escape, but air leakage around skylights can be a big problem, especially in colder climates.

The problem is not with the step-flashed skylight itself, but rather with the installation. Sealing a step-flashed skylight against air leakage requires felt paper to be run from the roof up the sides of the skylight frame, which means that the installer has to haul yet another item onto the roof in addition to all of the flashing pieces. If this step is skipped or not done properly, air from inside can escape through the spaces between the step flashing, along with lots of heating dollars. After a snowstorm here in Connecticut, it's common to see skylights on snow-covered roofs with halos of bare shingles (see the photo at left). In some cases melting can occur because of heat loss through the skylight frame, but air leakage is usually the culprit.

Skylights with solid flanges don't require felt-paper seals. The folks at Wasco do a little demonstration where they place a $50 bill under one of their fixed E-Class skylights on top of a solid table. Anyone who can lift the skylight off the table can have the bill. But because the flexible flange forms an airtight seal, the skylight won't budge, and no one has won the $50 yet. The flange functions the same way on the roof, forming an airtight seal against the roof sheathing and preventing warm air from finding its way out and around the skylight.

Bare shingles can indicate air leakage. Step-flashed skylights have to be sealed with felt paper under the flashing to make them airtight. If not done properly or skipped altogether, warm air leaking from inside will melt the snow around the perimeter of the skylight.

A shiny alternative finish. Sun-Tek offers its Classic Series skylights in polished copper with a clear, protective finish designed to keep the skylight shiny for years.

Designer Skylights Are Available in Colors

Most companies offer the choice of just a couple of colors. Roto, however, offers its Sunrise II skylights in five different colors including forest green and fire red. Other companies will custom-paint skylights to match or complement any funky color you might have on your roof. A word to the wise: Check out price and lead time before ordering your sea foam green skylights. You may decide that basic brown won't look so bad after all.

To me, the neatest-looking skylights are the copper-clad skylights available from Velux and Sun-Tek. Velux's copper-clad skylight is unfinished so that it will weather to a green patina. The Velux copper-clad skylight may be the best choice when skylights are used on historic buildings.

Sun-Tek's copper-clad skylight has a clear, protective finish for a shiny copper-kettle look (see the photo above). In the right application, this skylight has my vote for sexiest skylight on the market. The cost of upgrading to a copper-clad skylight is reasonable, adding $61 to the price of Velux's VS 106 and $33 to the price of Sun-Tek's VCG 2246.

One final word about availability. Most lumber stores carry many different makes and models and can order what you need if they don't have it in stock. If you are stuck for time and need an out-of-stock skylight yesterday, go with Velux. They pride themselves on being able to ship any of their standard skylights or accessories to your dealer free of charge within 24 hours of receiving an order.

Price estimates noted are from 1996.

Roe A. Osborn is a senior editor at Fine Homebuilding magazine.

Framing for Skylights

■ BY DOUG HOPPER

Figure out the thickness. The author uses a scrap of ½-in. plywood to determine the thickness of the furring he will nail to the headers at the bottom of the well.

Winter days in the Pacific Northwest can be short and gray, which helps explain why many people rely on skylights to make the most of what little natural light there is. When I started building 20 years ago, *skylight* wasn't even in my construction vocabulary. But skylights are now standard in the houses I build and are one of the improvements homeowners ask for most frequently in remodeling projects.

Framing for a skylight isn't difficult, but it's more than just cutting a hole in the roof and dropping in a factory-made unit. Installation is a lot simpler in new construction because roof and ceiling framing are open and visible, and the skylight can be included in framing plans right from the start. In a retrofit, you'll have to grapple with obstructions like heating ducts, plumbing, wiring and insulation, and none of these obstacles may be apparent when you start. Whether

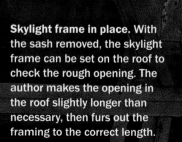

Skylight frame in place. With the sash removed, the skylight frame can be set on the roof to check the rough opening. The author makes the opening in the roof slightly longer than necessary, then furs out the framing to the correct length.

Making a Skylight Well with Plywood

Sloped lightwells are open to many variations. They can be sloped on four sides, sloped on one or two sides only, or even sloped more on one side than on another. I've found the easiest way to build the well, especially when it is irregular, is to use sheets of ¾-in. plywood (see the drawings below) rather than make a 2x frame.

I frame the rough opening in the ceiling ¾ in. larger all the way around. Then, on the underside of the rafter opening, I nail a 2x back from the edge of the end headers and side rafters by at least ¾ in. When the shaft angle is steep, I nail the 2x farther back from the edge so that the plywood sheets will line up with the edge of the roof opening. The steeper the angle of the well, the more setback will be required in the 2x nailer.

Next I cut sheets of plywood that fit from joists to rafters and nail them in place. This is a good place to use scraps. If the shaft is long, more than 3 ft. or 4 ft., I nail a 2x4 band around the outside of the well to stiffen the plywood structure. Drywall can be nailed or screwed directly to the plywood on the inside, and insulation stapled to the back.

An easy alternative to using 2x nailers is to cut a simple rabbet in the bottom of the rafters and the headers before they are installed. Plywood used for the walls of the well fit into the rabbet. I cut the rabbet with a circular saw and make it at least ¾ in. deep by about 1 in. wide. Make sure the wood is supported and held firmly in place. Once the rabbeted members are in place, the edges of the ¾-in. sheets can be nailed snugly into place.

Larry Haun is a carpenter in Los Angeles, California, and a frequent contributor to Fine Homebuilding.

Another method of making the well is to use plywood instead of 2x framing members. At the bottom of the well, the plywood is nailed to the inside of the headers and joists. At the top of the well, the plywood can be attached to a 2x nailer (drawing left) set back from the edge of the header or set into a rabbet (drawing right). Rabbets also can be cut into rafters and joists to accept the plywood walls on the side of the well.

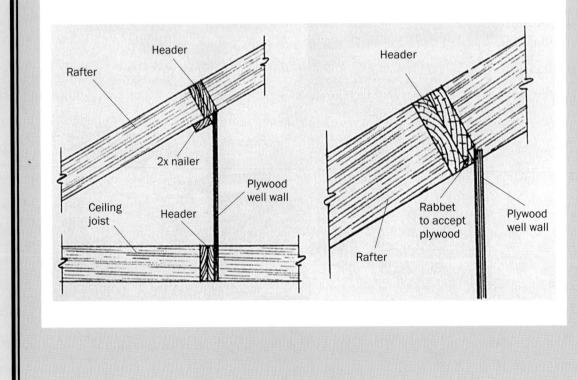

you're installing a skylight in a new house or an old one, though, framing is essentially the same. There are two basic framing questions you'll have to settle: how to create the openings in the roof and in the ceiling, and how to build the light shaft, or lightwell, that connects the two.

The size of the skylight has a lot to do with how complicated the framing will be. Some skylights fit between rafters or trusses on standard 24-in. on center spacing. If you choose one of these units, no rafters or ceiling joists must be cut, and framing is fairly simple. Large skylights require more work and a slightly different framing technique to make the openings and to build the lightwell. I'll explain how I handle both situations.

Planning the Skylight Well

What makes framing for a skylight unique is the lightwell, which connects openings in the roof and the ceiling. Of course, if there is no ceiling, such as in an open-beam roof, there is no well, and the only required framing will be at the roof. I frame my skylight wells with 2x material so that they can be insulated like an exterior wall (for tips on making well walls from plywood, see the sidebar on the facing page).

The size of the roof opening is determined by the rough opening of the skylight, but the size of the ceiling opening, and consequently the design of the well, is variable. A simple approach, and the one I use when the well will not be very deep (less than 2 ft. high at the high end), is to bring all four sides of the well down 90° to the roof slope. More often, though, I run the high end of the well 90° to the roof slope, and the lower end of the well plumb. The ceiling opening then becomes longer than the roof opening but the same width as the skylight. The splayed shape allows for greater dispersion of light inside.

This approach works well unless the depth of the well is more than about 4 ft. at the high end. With very deep wells, the ceiling opening can get too long if the high end is run 90° to the roof. An obvious solution is to reduce the angle so that the opening is somewhat shorter or to bevel the high-end wall. A beveled wall starts perpendicular to the roof slope and then breaks about halfway down to continue plumb, or perpendicular, to the ceiling. But before I build that kind of well, I first try to adjust the angles of the well because I think the finished well looks nicer when the walls are each in a single plane.

Another option is to make the ceiling opening greater in width and length than the roof opening. This distributes the light more than the simple splayed-end well, but it's more complicated to build because of the compound angles involved. In addition, it usually means more structural work at the ceiling because you probably will have to cut more joists to make the opening. For those reasons, I don't use this approach very often. Yet it's worth considering, especially when using a small skylight because the distribution of light is increased greatly.

Locating the Well Inside

In new construction, I often start my skylight framing on the floor below the well, not on the roof. After all of the exterior and interior walls have been laid out on the floor deck, I lay out approximate dimensions for the well on the deck. It may sound backward, but starting the process on the floor lets me see how the skylight well will fit with other openings in the walls, the roof and the ceiling. Sometimes I know how big I want the well to be at the ceiling, but I'm not sure how big a skylight to order. If that's the case, I can use a simple formula to make

Framing for a skylight isn't difficult, but it's more than just cutting a hole in the roof and dropping in a factory-made unit.

Sizing a Skylight from Below

Because I usually build skylight wells with a splayed wall at the high end, the opening in the ceiling will be larger than the rough opening at the roof. What if you know how big the well should be at the ceiling, but you're not sure what size skylight to order? A simple formula can help you solve this layout and framing problem.

Let's say your goal is to frame a 5-ft.-long well opening in the middle of a 10-ft.-wide ceiling, leaving 2 ft. 6 in. on both sides of the opening (see the drawing below). Your object is to figure out the rough opening on the roof, assuming the high end of the well is perpendicular to the roof plane, and the low end of the well is plumb. The rough opening will tell you what size skylight to order. The same formula could be used to work the other way, that is, starting with the rough opening and figuring out the size of the well opening in the ceiling.

Here's the formula for figuring out the skylight problem: $(B \div 12) \times \text{slope factor} = A$

A1 in the drawing is the horizontal distance from the outside edge of the rafter to the exterior wall (10½ in. in this example), the width of the 2x6 stud

in the exterior wall (5½ in.) plus our desired well opening (60 in.) plus the setback from the outside wall (30 in.) for a total of 106 in. B2 in the drawing is 46 in. We also need to know the slope factor, which is the hypotenuse of the triangle formed by the roof. This 9-in-12 roof has a slope factor of 15. (You can get this number right off your framing square in the line named "length of common rafters per foot run.")

It's not hard to plug these values into the formula and get the answers. To determine A2, you use the formula as shown: $(46 \div 12) \times 15 = 57\frac{1}{2}$ in. To determine B1, you have to invert the formula because the known quantity, A1, is the hypotenuse of the larger right triangle, not the long leg as in the smaller triangle. So, $(A \div \text{slope}) \times 12 = B$ or $(106 \div 15) = 84\frac{13}{16}$ in. The difference between B1 and A2, $27\frac{5}{16}$ in., is the length of the rough opening. The width of the skylight can vary and doesn't really affect this layout.

With these numbers in hand, I can order the skylight that comes closest to my design goals. I'd wait until I had the unit on the site before I framed the openings and built the well.

Laying Out the Well and Roof Opening

Framing for a skylight usually requires two different size openings—one in the roof and one in the ceiling. You may start by knowing the size of one of them but not the other. If that's the case, a simple formula can help you calculate the size of the second opening and complete the layout for framing.

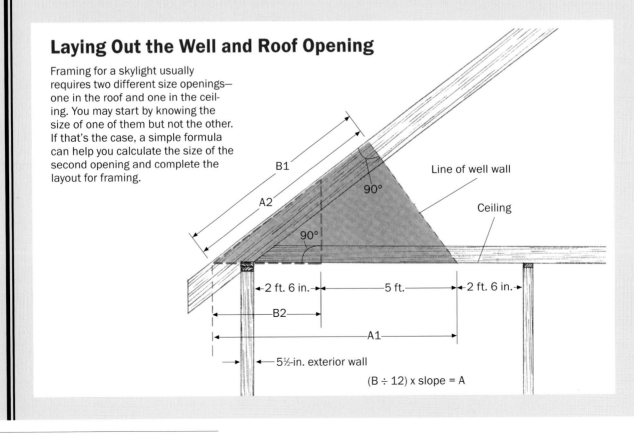

$(B \div 12) \times \text{slope} = A$

the calculation (see "Sizing a Skylight from Below" on the facing page).

Laying out the well opening on the floor also gives me a chance to make minor adjustments in ceiling and roof framing before it goes up. For instance, a common location for a skylight is centered over the kitchen sink and window. In the case of a single-bay skylight, I may be able to shift roof and ceiling framing a few inches in either direction so that the framing won't encroach on the skylight well. This method is a lot faster than framing everything and then going back later to frame in the roof opening—especially when the roof opening wouldn't require headers if located properly. A potential drawback is when a shift in rafter or joist spacing means there will be too great a span for roof sheathing or material I'm going to use on the ceiling. If that's the case, I just drop in an extra rafter or joist to keep spans within allowable limits.

In a retrofit, it may be easier to start the layout at the ceiling. This is especially true when you're trying to avoid an obstruction inside, like a wall, and the location of the skylight on the roof isn't critical. But be careful to avoid vents, roof valleys and other obstructions.

Framing on the Roof

Once I know where the well is going, I plumb up from the floor to the ceiling joists and mark one end of the well wall. Those marks can then be transferred to the roof plane, and the roof opening marked on the rafters. I usually add a little bit to the length of the rough opening, 1 in. to 1½ in., and I'll tell you why a little later. The width of the

---

Framing for Single-Bay Skylights

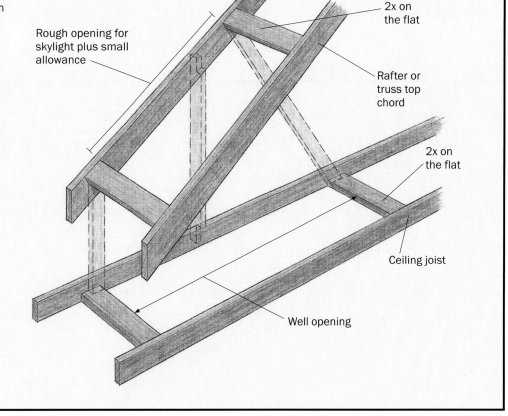

When a skylight will fit between rafters or roof trusses, the top and bottom plates of the well walls can be 2xs nailed on the flat. These 2xs won't carry any loads from interrupted roof or ceiling members and must support only the weight of the finish material inside the well. The openings in the roof and the ceiling are connected with studs the same width as the wall framing. That allows the well to be insulated just like exterior walls. The high-end wall of the well is 90° to the roof slope; the low end is plumb.

Rough opening for skylight plus small allowance

2x on the flat

Rafter or truss top chord

2x on the flat

Ceiling joist

Well opening

Make way for a skylight. If a skylight can't fit between framing members, a rafter will have to be cut back. Establish the cutline by holding a framing square to the inside of the adjacent rafter and marking the rafter to be cut.

opening should be exactly what the skylight manufacturer specifies. I'll complete the ceiling and well-wall framing from the inside a little later, but once I've marked the rough opening for the skylight on the rafters, I'm ready to frame the skylight opening at the roofline. This is the first step, whether it's new construction or a retrofit.

How I frame the opening depends on whether the skylight will fit between framing members on standard spacing. If the size of the skylight won't require that any rafters be cut, I use single framing members laid flat (parallel to the roof slope) between adjacent rafters to serve as the top plates for the end well walls that I'll build shortly (see the drawing on p. 127). If it's necessary to cut structural members and head off the framing (see the photo above), the standard practice is to double the adjacent framing members

(rafters and ceiling joists) that are not cut, and double and set on edge the cross framing, or headers (see the drawing on the facing page).

You should use metal hangers on all connections because the headers are now carrying the load of the interrupted joists or rafters (see the left photo on the facing page). If you have to head off an interrupted rafter, the opening may be too large for the skylight. If so, a single rafter between the headers will narrow the opening to the right size (see the right photo on the facing page). As a general rule, if I must head off more than one joist or rafter, I'll have the design checked by an engineer.

Finally, the roof sheathing goes on right across the opening. I don't cut the sheathing until I'm ready to install the skylight. That means the job site will be safer because there's one less hole on the roof to fall through, and the house will be less prone to weather damage.

With the rough opening framed in the roof, I can plumb down to the ceiling joists at the low end of the well and mark the location for the wall there (see the drawing on p. 132). The wall at the high end of the well is perpendicular to the roof, and I use a framing square and a straightedge from the roof to locate the inside of the well wall on the ceiling joists. The ceiling opening can then be framed in the same way the roof opening was.

Framing the Well

I install the four corner studs of the well after the plates or headers have been nailed between the rafters and the joists. In my part of the country, energy codes require that skylight wells be insulated, so I use the same size framing members for the well as I do for the wall studs (usually 2x6s). For single-bay skylights, with top and bottom plates of the well walls made from 2xs on the flat, the easiest way to measure for the well corner pieces is to

Framing Wider Skylights

When a skylight is too wide to fit between rafters and joists on standard spacing, structural members must be cut and headed off. Headers pick up the load from the interrupted rafters or joists, so connections between headers, joists and rafters should be reinforced with metal hangers. The author doubles up both headers and framing members on both sides of the opening. Studs that form the high end of the well, perpendicular to the roof plane, are notched around the header.

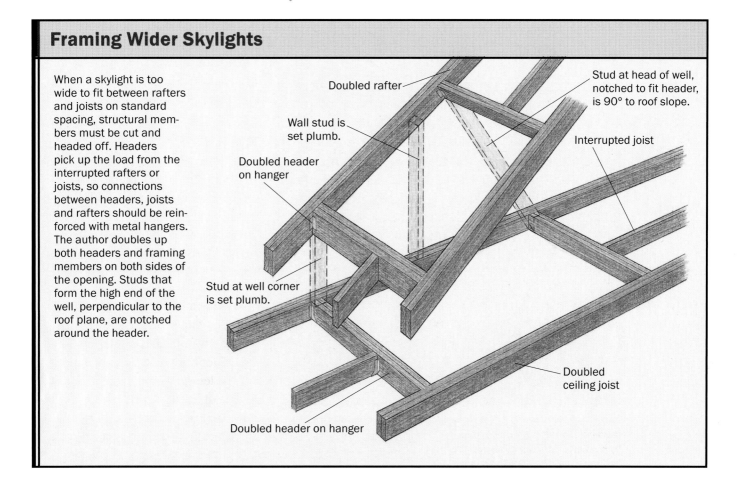

Doubled rafter

Stud at head of well, notched to fit header, is 90° to roof slope.

Wall stud is set plumb.

Interrupted joist

Doubled header on hanger

Stud at well corner is set plumb.

Doubled ceiling joist

Doubled header on hanger

Metal hangers for a strong connection. Because the header is supporting part of the roof load, the author uses a double 2x on edge and connects the header to adjacent rafters with metal hangers.

Establishing the well width. Using a short section of a 2x4, the author wedges a rafter in place and toenails it to the header. He'll add a metal hanger later to help carry the roof loads.

Retrofitting Skylights in a Truss Roof

Reroofing is a great time to install skylights. However, a truss roof can make skylight installation a lot more problematic. Building codes prohibit any field modification (notching or cutting) of roof trusses. That's because once you cut into any part of a roof truss, you can't trust its integrity. So you really should check with your local building inspector before proceeding. But the easiest way I've found to retrofit a wide skylight without re-engineering the trusses is to reframe the roof around the skylight opening with dimensional lumber just as you would in a stick-built roof.

First, you have to plan the location for the skylight. Because 4-ft. skylights are sized to fit pretty closely between roof framing spaced 24 in. on center, you may have to cut through one or two roof trusses to frame out the opening. I actually prefer to locate skylights so that their sides fall in the middle of truss bays and not alongside existing trusses. This strategy gives a little more room for slipping in the new rafters and ceiling joists.

Next, calculate the size of the rafters and ceiling joists that will support the roof where you cut the trusses. I use building code books or span tables to determine rafter size and spacing. Before you get too far into the tables, measure the distance between the inside edge of the top wall plate and the underside of the roof sheathing. This amount is the limiting height you will have to work with when selecting an appropriate rafter size.

You should not extend the heel cut on the rafters too far beyond the inside edge of the top plate. Usually, roof-truss height at the top plate limits the rafter sizing to 2x8s or 2x6s. Ceiling joists can

be a little taller. Selecting a premium species such as Douglas fir or a premium grade such as select structural might be necessary to get away with 2x6 rafters on longer spans.

If you cannot support the roof properly, I suggest going with two 2-ft. skylights or leaving a truss exposed in the skylight shaft. But let's assume you can size the rafters and ceiling joists and move on to the next step of cutting and installing them.

If possible, I remove several sheets of roof sheathing in the area where the new skylight will go, which gives you plenty of room to do the framing. If this option isn't possible, the lumber can be slipped through small holes cut through the sheathing at the ridge, and the framing can be done from inside the roof cavity.

Cut enough rafters for both sides of the roof. Even though you'll be putting the rough opening for the skylight only on one side, you need to reframe both sides of the roof. Double the rafters on each side of the rough opening, and prepare headers and cripple rafters.

Don't cut any roof trusses until you get most of the new framing in place. Without getting too involved in describing the logistics, slide in the rafters, and install them first. When the rafters are in position, hop out on the roof and drive nails through the roof sheathing into the rafters wherever you can.

Next, install the ceiling joists. Then prepare a ridge board that will fit between the roof trusses that will remain intact after you cut the trusses for the skylight opening. Using a reciprocating saw, cut out sections of the top chord of the one or two trusses in the way so that you can slip the ridge board in place. A metal-cutting blade will be necessary to cut through the truss plates at the top of the trusses. By the way, this ridge is not structural. It's merely there to help transfer opposing rafter forces.

With the ridge in place, the rafters are supporting the roof, and you can cut out sections of roof trusses in the way of the new skylight opening. The rest of the truss can remain. In fact, some parts of the truss have to remain to support the soffit and fascia.

Next, determine the exact location of the ceiling opening for the skylight, and frame in the headers and joists. Mark out the skylight opening on the underside of the roof sheathing, and cut the top chords of the trusses just short enough to allow you to install the headers and the cripple rafters. Be sure to secure all rafter–header connections with metal connectors. Do the same for the remaining segments of the roof-truss chords. Then frame up the skylight shaft. If you're not ready to reroof, schedule the final cut in the roof sheathing and skylight installation just before the new roof is installed.

Mike Guertin, *a contributing editor to Fine Homebuilding, lives and works in East Greenwich, Rhode Island. He is the co-author of* Precision Framing *(The Taunton Press, 2001), and the author of* Roofing with Asphalt Shingles *(The Taunton Press, 2002).*

New Rafters Replace Trusses around Skylight

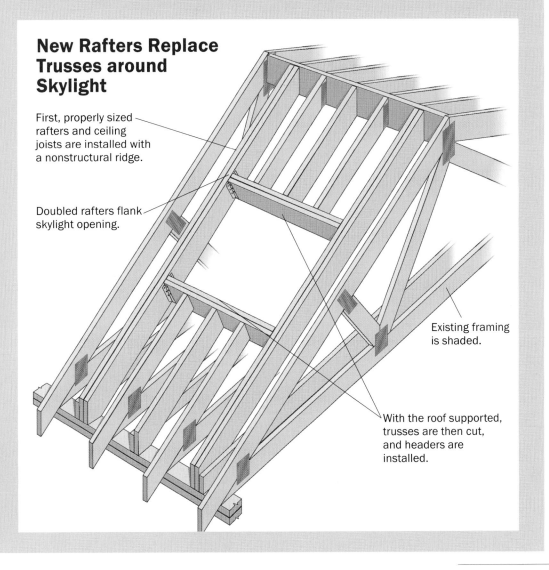

First, properly sized rafters and ceiling joists are installed with a nonstructural ridge.

Doubled rafters flank skylight opening.

Existing framing is shaded.

With the roof supported, trusses are then cut, and headers are installed.

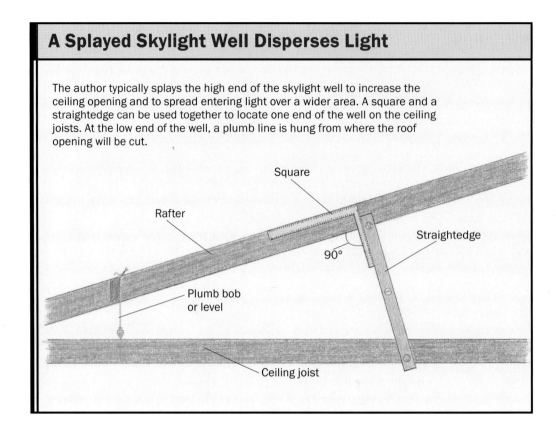

A Splayed Skylight Well Disperses Light

The author typically splays the high end of the skylight well to increase the ceiling opening and to spread entering light over a wider area. A square and a straightedge can be used together to locate one end of the well on the ceiling joists. At the low end of the well, a plumb line is hung from where the roof opening will be cut.

Square

Rafter

Straightedge

90°

Plumb bob
or level

Ceiling joist

cut the top end of the studs at 90° or at the slope of the roof, depending on which end of the well I'm working on. I make the studs a couple of inches longer than their finished lengths. With locations for bottom plates marked on the ceiling joists, but the plates not yet installed, I hold the corner pieces in place and mark them at the bottom edge of the ceiling joist to get an exact measurement for length. Before cutting, I deduct 1½ in. for the thickness of the bottom plates.

Using the same technique for determining length and angle, I fill in the side and end framing, spacing the studs just like those in the wall framing (for 2x6s, that would be 24 in. on center). On the well's sidewalls, I usually have only a single rafter or joist at top and bottom to which I nail the studs, so I notch the studs over the rafters and the joists to get better nailing.

Framing the well is a little different when structural members, either rafters or ceiling joists, are cut to accommodate the skylight.

In that instance, the 2x headers will be on edge. At the high end of the well, the tops of the studs are cut square. The bottom of the studs are notched around the header so that the wall of the well will plane smoothly into the ceiling. The wall at the lower end is framed just the opposite way, with square cuts at the bottom and angle cuts at the top.

After the well has been framed, I drive nails up through the roof at the corners from the inside, mark the dimensions of the opening on the roof, and then cut the opening.

Making Final Adjustments in the Well

Here's why I usually allow a little extra room in the length of the well: The wall on the low end of the well is plumb and therefore intersects the roof plane at an angle. That makes it a little tricky to predict where the line of the wall will emerge through the roof sheathing. It's easier to make the finish opening a little smaller after it's framed than it is to enlarge it once all the framing is in place. So I shim the lower end of the wall out after I've put the skylight on the roof and can measure exactly how much space I have to take up.

I go through this trouble so that the finish material I use on the well lines up with the appropriate spot on the skylight frame. Some skylights have grooves cut in the bottom edges of the frame. These grooves accept the finish material on the sides of the well and make a clean finish. There isn't much room for error in the well framing. After the roof opening has been cut, I remove the sash from the skylight, put the frame on the roof, and check the framing (see the photo on p. 123). With a fixed unit, this step may require two people: one outside and one inside. If I'm using ½-in. drywall on the inside of the well, for instance, I'll use a scrap of ½-in. plywood as a gauge to align the well framing with the skylight frame (see the photo on p. 122). At first glance this may not seem necessary, but because of the angle at the bottom of the well, it can be difficult to align the wall solely with a tape or by eye.

Framing in a Truss Roof

Framing in a skylight on a truss roof shouldn't be a problem as long as the skylight fits between the typical 24-in. on center spacing of the trusses. Because you can't modify trusses in the field without consulting an engineer, a skylight that's too big for this opening can be a problem.

I first would use two or more smaller units that would fit within standard framing. The skylights can be ganged together, either side by side or top and bottom, with manufactured flashing designed to span typical framing-member widths. If you place the units side by side, you will build two identical wells separated by a truss.

In new construction, another option is to plan the truss design to allow for a larger opening. This typically means leaving out one truss to create a nominal 4-ft. opening. Be sure to notify the truss manufacturer so that appropriate adjustments can be made to the design.

If I'm leaving out a long-span truss for a skylight (and the plan has been okayed by an engineer or the architect), I'll fill the area between full trusses with what I call ladder framing. This consists of 2xs nailed on edge with hangers at each end laid out to align with my sheathing courses.

Doug Hopper is a builder in Tacoma, Washington.

CREDITS

p. iii: Photo by Tony Simmonds.

p. iv: (left) Photo by Roe A. Osborn, courtesy of *Fine Homebuilding,* © The Taunton Press, Inc.; (right) Photo by Charles Miller, courtesy of *Fine Homebuilding,* © The Taunton Press, Inc.

p. v: Photos by Roe A. Osborn, courtesy of *Fine Homebuilding,* © The Taunton Press, Inc.

The articles in this book appeared in the following issues of *Fine Homebuilding:*

p. 2: Airtight Attic Access by Mike Guertin, issue 148. Photos by Roe A. Osborn, courtesy of *Fine Homebuilding,* © The Taunton Press, Inc., except photos on p. 4 (left, top and bottom) by Tom O'Brien, courtesy of *Fine Homebuilding,* © The Taunton Press, Inc.; Drawings by Don Mannes.

p. 6: Disappearing Attic Stairways by William T. Cox, issue 89. Photos by Jefferson Kolle, courtesy of *Fine Homebuilding,* © The Taunton Press, Inc., except photo on p. 10 (bottom) by William T. Cox.

p. 14: Fixing a Cold, Drafty House by Fred Lugano, issue 105. Photos by Andrew Wormer, courtesy of *Fine Homebuilding,* © The Taunton Press, Inc., except photo on p. 17 by Fred Lugano and photo on p. 18 by Scott Phillips, courtesy of *Fine Homebuilding,* © The Taunton Press, Inc.; Drawing by Mark Hannon.

p. 25: Bed Alcove by Tony Simmonds, issue 76. Photos by Tony Simmonds, except photo on p. 25 by Charles Miller, courtesy of *Fine Homebuilding,* © The Taunton Press, Inc.; Drawings by Bob LaPointe.

p. 33: A Fresh Look for an Attic Bath by Jack Burnett-Shaw and Julia Strickland, issue 135. Photos by Roe A. Osborn, courtesy of *Fine Homebuilding,* © The Taunton Press, Inc.; Drawing by Paul Perreault.

p. 38: Adding On, but Staying Small by Harry N. Pharr, issue 120. Photos by Charles Miller, courtesy of *Fine Homebuilding,* © The Taunton Press, Inc.; Drawings on p. 40 (top) by Harry N. Pharr, p. 40 (bottom) and p. 42 by Scott Bricher.

p. 46: Jewelbox Bathroom by Jeff Morse, issue 76. Photos by Charles Miller, courtesy of *Fine Home-building,* © The Taunton Press, Inc.; Drawings by Gary Williamson.

p. 54: Adding a Second Story by Tony Simmonds, issue 102. Photos by Tony Simmonds, except photos on pp. 59, 60, 61, and 62 by Charles Miller, courtesy of *Fine Homebuilding,* © The Taunton Press, Inc.; Drawings by Bob LaPointe.

p. 64: A Gable-Dormer Retrofit by Scott McBride, issue 134. Photos by Scott McBride; Drawings by Christopher Clapp.

p. 74: Framing an Elegant Dormer by John Spier, issue 130. Photos by Roe A. Osborn, courtesy of *Fine Homebuilding,* © The Taunton Press, Inc.; Drawings by Christopher Clapp.

p. 84: Keeping a Dormer Addition Clean and Dry by Nicholas Pitz, issue 158. Photos by Nicholas Pitz; Drawings by Don Mannes.

p. 90: Framing a Dramatic Dormer by John Spier, issue 150. Photos by Roe A. Osborn, courtesy of *Fine Homebuilding,* © The Taunton Press, Inc.; Drawings by Bob LaPointe.

p. 97: Dramatic Skylight by Mike Guertin, issue 164. Photos by Roe A. Osborn, courtesy of *Fine Homebuilding,* © The Taunton Press, Inc.; Drawings by Dan Thornton.

p. 105: Skylight Kitchen by George Burman and Patricia Looney-Burman, issue 143. Photos by Charles Miller, courtesy of *Fine Homebuilding,* © The Taunton Press, Inc., except photo on p. 107 (top) by Scott Phillips, courtesy of *Fine Homebuilding,* © The Taunton Press, Inc.; Drawings by Paul Perreault.

p. 111: Shedding Light on Skylights by Roe A. Osborn, issue 102. Photos by Roe A. Osborn, courtesy of *Fine Homebuilding,* © The Taunton Press, Inc., except photo on p. 121 courtesy Sun-Tek, and photos on pp. 113, 114, and 117 (bottom) courtesy Velux-America; Drawings by Christopher Clapp.

p. 122: Framing for Skylights by Doug Hopper, issue 90. Photos by Scott Gibson, courtesy of *Fine Homebuilding,* © The Taunton Press, Inc.; Drawings by Vince Babak.

INDEX

Taunton's FOR PROS BY PROS Series
A collection of the best articles from *Fine Homebuilding* magazine.

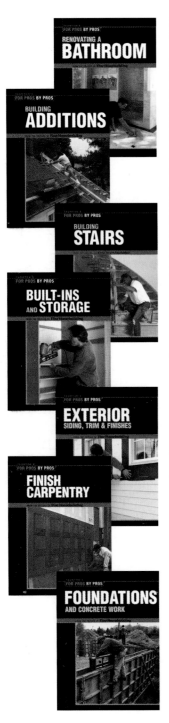

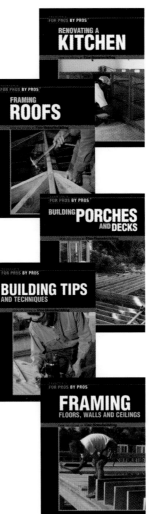

Other Books in the Series:

Taunton's For Pros By Pros:
RENOVATING A BATHROOM

ISBN 1-56158-584-X
Product #070702
$17.95 U.S.
$25.95 Canada

Taunton's For Pros By Pros:
BUILDING ADDITIONS

ISBN 1-56158-699-4
Product #070779
$17.95 U.S.
$25.95 Canada

Taunton's For Pros By Pros:
BUILDING STAIRS

ISBN 1-56158-653-6
Product #070742
$17.95 U.S.
$25.95 Canada

Taunton's For Pros By Pros:
BUILT-INS AND STORAGE

ISBN 1-56158-700-1
Product #070780
$17.95 U.S.
$25.95 Canada

Taunton's For Pros By Pros:
EXTERIOR SIDING,
TRIM & FINISHES

ISBN 1-56158-652-8
Product #070741
$17.95 U.S.
$25.95 Canada

Taunton's For Pros By Pros:
FINISH CARPENTRY

ISBN 1-56158-536-X
Product #070633
$17.95 U.S.
$25.95 Canada

Taunton's For Pros By Pros:
FOUNDATIONS AND
CONCRETE WORK

ISBN 1-56158-537-8
Product #070635
$17.95 U.S.
$25.95 Canada

Taunton's For Pros By Pros:
RENOVATING A KITCHEN

ISBN 1-56158-540-8
Product #070637
$17.95 U.S.
$25.95 Canada

Taunton's For Pros By Pros:
FRAMING ROOFS

ISBN 1-56158-538-6
Product #070634
$17.95 U.S.
$25.95 Canada

Taunton's For Pros By Pros:
BUILDING PORCHES AND
DECKS

ISBN 1-56158-539-4
Product #070636
$17.95 U.S.
$25.95 Canada

Taunton's For Pros By Pros:
BUILDING TIPS AND
TECHNIQUES

ISBN 1-56158-687-0
Product #070766
$17.95 U.S.
$25.95 Canada

Taunton's For Pros By Pros:
FRAMING FLOORS,
WALLS, AND CEILINGS

ISBN 1-56158-758-3
Product #070821
$17.95 U.S.
$25.95 Canada

For more information visit our website at www.taunton.com.